Lecture Notes in Mathematics

Edited by A. Dold, F. Takens and B. Teissier

Editorial Policy
for the publication of monographs

1. Lecture Notes aim to report new developments in all areas of mathematics – quickly, informally and at a high level. Monograph manuscripts should be reasonably self-contained and rounded off. Thus they may, and often will, present not only results of the author but also related work by other people. They may be based on specialized lecture courses. Furthermore, the manuscripts should provide sufficient motivation, examples and applications. This clearly distinguishes Lecture Notes from journal articles or technical reports which normally are very concise. Articles intended for a journal but too long to be accepted by most journals, usually do not have this "lecture notes" character. For similar reasons it is unusual for doctoral theses to be accepted for the Lecture Notes series.

2. Manuscripts should be submitted (preferably in duplicate) either to one of the series editors or to Springer-Verlag, Heidelberg. In general, manuscripts will be sent out to 2 external referees for evaluation. If a decision cannot yet be reached on the basis of the first 2 reports, further referees may be contacted: the author will be informed of this. A final decision to publish can be made only on the basis of the complete manuscript, however a refereeing process leading to a preliminary decision can be based on a pre-final or incomplete manuscript. The strict minimum amount of material that will be considered should include a detailed outline describing the planned contents of each chapter, a bibliography and several sample chapters.
Authors should be aware that incomplete or insufficiently close to final manuscripts almost always result in longer refereeing times and nevertheless unclear referees' recommendations, making further refereeing of a final draft necessary.
Authors should also be aware that parallel submission of their manuscript to another publisher while under consideration for LNM will in general lead to immediate rejection.

3. Manuscripts should in general be submitted in English.
Final manuscripts should contain at least 100 pages of mathematical text and should include
– a table of contents;
– an informative introduction, with adequate motivation and perhaps some historical remarks: it should be accessible to a reader not intimately familiar with the topic treated;
– a subject index: as a rule this is genuinely helpful for the reader.

Continued on back inside cover

Lecture Notes in Mathematics

1525

Editors:
A. Dold, Heidelberg
B. Eckmann, Zürich
F. Takens, Groningen

Mathematical Research Today and Tomorrow

Viewpoints of Seven Fields Medalists

Lectures given at the Institut d'Estudis Catalans,
Barcelona, Spain, June 1991

Editors: C. Casacuberta, M. Castellet

Springer-Verlag
Berlin Heidelberg New York
London Paris Tokyo
Hong Kong Barcelona
Budapest

Editors

Carles Casacuberta
Departament d'Àlgebra i Geometria
Universitat de Barcelona
Gran Via de les Corts Catalanes, 585
E-08007 Barcelona, Spain

Manuel Castellet
Departament de Matemàtiques
Universitat Autònoma de Barcelona
E-08193 Bellaterra, Spain

Cover drawing by Joan Solà.

Photo of Sergei Novikov by Jordi Morera; courtesy of *El Temps*.

Photos of Alain Connes, Gerd Faltings, Vaughan Jones, Stephen Smale and René Thom by Salvador Esparbé; courtesy of the Science Museum of Barcelona.

Mathematics Subject Classification (1991): 00-02, 00B20

ISBN 3-540-56011-4 Springer-Verlag Berlin Heidelberg New York
ISBN 0-387-56011-4 Springer-Verlag New York Berlin Heidelberg

Springer-Verlag Berlin Heidelberg New York
a member of BertelsmannSpringer Science+Business Media GmbH

© Springer-Verlag Berlin Heidelberg 1992
Printed in Germany

Typesetting: Camera-ready by author/editor
41/3111-5432 - Printed on acid-free paper

Preface

As Beno Eckmann said in his Inaugural Lecture for the academic year 1978–79 at the ETH Zürich, "Nowadays, mathematics plays three different rôles: It is an important and essential tool in our modern scientific and technological world, it is a model to seek the objective truth in a perfectly defined context, and, moreover, forms part of the basis of our cultural tradition."

From 13th to 18th of June 1991, the *Institut d'Estudis Catalans* —an academic and cultural body whose aim is higher scientific research and, in particular, research into all elements of Catalan culture— presented to the mathematical community the *Symposium on the Current State and Prospects of Mathematics*, whose main goal was to combine the above three aspects.

Among the different cultural events preceding the 1992 Barcelona Olympic Games, mathematics was expected to play —and in fact has played— an outstanding rôle. Thanks to the financial support coming not only from the *Olimpíada Cultural* but also from the CIRIT and DGICYT, the *Centre de Recerca Matemàtica* (CRM) was able to gather seven Fields Medalists in Barcelona, who presented the most recent advances and debated the present and future of mathematical research.

"Mathematics is a truly international science, perhaps the most international of sciences, since, compared with other disciplines, it is based less on the use of instruments and tools and more on human contact," said Peter Hilton on the occasion of the 20th anniversary of the FIM Zürich. It is precisely in this sense that research institutes play an essential rôle, facilitating not only the exchange of ideas between specialists in the same fields, but also favouring the growth of deep-rooted and sometimes surprising relations between different lines of research. This is the true purpose of the CRM: to be a real laboratory for mathematical contacts.

The present volume contains the text of the seven invited lectures by Fields Medalists, as well as a faithful transcription of the round-table discussion which was held at the end of the Symposium.

Shing-Tung Yau, who was unable to come to Barcelona, kindly submitted the text of his lecture. We extend our warm thanks to him, and to all the other authors for their contributions. We are indebted to our friends and colleagues who have collaborated in the production of this volume; their names are mentioned at the end of the corresponding articles.

Table of Contents

Symposium on the Current State
and Prospects of Mathematics

Barcelona, June 1991

Leaving Mathematics for Philosophy

by

René Thom

Fields Medal 1958

for inventing and developing the theory of cobordism in algebraic topology. This classification of manifolds used homotopy theory in a fundamental way and became a prime example of a general cohomology theory.

Abstract: This lecture is an overview of the author's own work from homotopy theory and homological algebra to the recent books *Esquisse d'une Sémiophysique* and *Apologie du Logos*, with emphasis on the rôle played by different ideas from biology in catastrophe theory.

The attitude adopted is critical towards experimentation with no underlying theory and towards an improper use of statistics as a justification of analogical thinking in science.

Leaving Mathematics for Philosophy

I must say, to begin, how happy I am to be in Barcelona again. Visiting Barcelona, I think, is a kind of continuous surprise and it is something which is always what the Germans would call *ein Ereignis*.

My talk consists of what happened to me after being awarded the Fields Medal and how, finally, I survived this event.

When, a little while before the once fateful limit of 35 years, you receive the Fields Medal, when you have been subjected to the rituals of the International Congress of Mathematicians, where you get the medal, and, tired but happy, you are left to yourself again, it is then that you are assailed with doubts. Am I worthy of this honour done to me? Were there not to be found, amongst younger mathematicians, some who were more deserving? And, more subtle in its perversity, a suspicion creeps into your mind: Was not this the highest level of achievement I shall ever attain? Whatever I do later on, is it not bound to be something less? Will my mathematical capacities always stay at this level or am I doomed to an inevitable falling off?

All these thoughts were at the back of my mind already in the spring of 1958. Shortly before the date of the Congress came Barry Mazur's proof of the Schönflies conjecture —to my mind a pure gem of the deepest mathematical inventiveness. A little before that, Milnor's exotic sphere structures had appeared. (Milnor did later receive the Fields Medal, this is true.) It seemed more and more evident to me that my capacities for rigorous mathematical proof were lessening as time went by. It was then that I abandoned the more algebraic theories (like homotopy, homological algebra, K-theory, etc.) which bore little attraction for me, and turned again to study the singularities of differentiable maps, a subject to which Hassler Whitney had given a brilliant start in 1951–52, and which appeared to me more flexible and more concrete.

Singularity Theory

I then tried to find out whether the idea of "genericity" of a singularity (like the classical cusp for smooth mappings of one plane onto another) might have applications in everyday life. I remember as I lay on my berth on the liner *Île de France* bearing me to New York and Princeton during my first Atlantic crossing in 1951, that the gentle rocking of the swell gave me the idea of comparing the unfolding of a wave to the transformation of a smooth regular curve obtained as an image of a map $f : I \to \mathbf{R}^2$, which, passing through an ordinary cusp, develops a double point that subsequently breaks, liberating (by surface tension) an isolated "drop" of the initial wave (a phenomenon which can be easily observed in a linear pencil of cubic curves). So, I already had this general idea that looking at a very simple phenomenon of ordinary type, it is not impossible to find for it simple algebraic models for which one can find a sort of a priori generalization or justification.

At Strasbourg University, towards 1959–60, I began, with the help of a physicist colleague, to study the singularities of caustics in optical geometry, as well as their deformations. I was amazed to discover that caustics developed more singularities than required by the simple theory of transversality for plane curves. The singularities of curves in the plane are fairly well known and very easy to describe. To my surprise, I found that in caustics, organized by some very simple optical instruments like spherical mirrors and rectilinear diopters, one may find a singularity that should not theoretically exist. This was the case of the so-called *umbilic points*, which I will discuss later. It took me several years to understand that this was a consequence of Fermat's principle of optimality. An observer aware of this could have foreseen the existence of the optimality principle simply by playing around with caustics.

In 1963 I left Strasbourg for the IHES at Bures-sur-Yvette. There I enjoyed total freedom, with no teaching or administrative tasks imposed on me. About that time, the general idea of sets and stratified morphisms took shape in my mind. In particular, one year before (this was the beginning of the movement of modern mathematics), people wanted, in general, to drop geometry out of the curriculum. Especially the theory of envelopes, because the theory of envelopes was a rather difficult theory to explain and to make precise. So they decided to eliminate the theory of envelopes from the curriculum. This irritated me quite a lot, so I wrote an article trying to explain that the notion of envelopes was just a special case of the general notion of singularity of a map, and if we knew how to deal with generic singularities of maps, then we should also know about envelopes. Thinking about that, I developed essentially the theory of singularities; in particular, the extension of transversality theory to the special jets, which I had already made in 1956.

Then I tried, using Whitney's formulation of the (a, b) property, to get the extension of transversality theory to spaces satisfying, for the connection between the corresponding subsets which were later called *strata*, the corresponding property to the classical stability property of transversal intersection for smooth manifolds.

But the axiomatics I worked out were still inadequate, and it was not until John Mather's work, around 1966, that a satisfactory description of the chief properties of stratified objects and strata —in particular, the topological stability of almost smooth proper maps between manifolds— was proved. This work culminated in my article in the Bulletin of the American Mathematical Society in 1968, and was practically my last piece of strictly mathematical research. After that, I turned into —or "degenerated into"— a philosopher.

The Origins of Catastrophe Theory

This departure from mathematics towards philosophy did not happen at once. Around 1963–64, whilst undertaking the study of classical embryology, I pursued my reflections on the "concrete" use of transversality theorems in the direction of biology. In some sense, I got the feeling that, as a mathematician, I was quite replaceable; that anything I could find would certainly be found without me shortly after, and, in general, in a better way. As a result, I tried to get out of the general program of mathematical production and I wanted to study, precisely, the "concrete" use of transversality theory for natural forms. So in 1965–66 I started to work on what was to become, six years later, catastrophe theory. In these beginnings, the ideas of two biologists, very different one from the other, played an essential part. One was Conrad H. Waddington, with his model of the *epigenetic landscape*. The epigenetic landscape is a sort of landscaping hill with several valleys running down from the top of the hill. It was supposed to describe the succession of cellular differentiations in embryology, each valley describing a common fate of a clone of cells, and every valley bifurcating into two valleys. Following one branch or another would correspond to the choice of one cell between one type of cellular differentiation, and another cell following the other type of differentiation.

The other biologist who had more or less the same kind of idea was Max Delbrück in a paper presented —I think in 1946— in a congress of the CNRS in Paris, where he linked cellular differentiation with what is called a *stable mitotic regime* of embryo cells; that is, the idea that in an embryo dividing by mitosis there could be stable asymptotic regimes of mitosis, which could eventually bifurcate (locally) and create the corresponding clones.

It was in trying to apply this model to embryology that I ended up with my famous list of seven *elementary catastrophes* in space-time.

Meanwhile, singularity theory made considerable headway, already in 1960, with Malgrange's proof of the C^∞ preparation theorem, the use of transversality in stratified set theory, and later (1968–69) with the explanation by Arnol'd and his Moscow school of the general theory of singularities of analytic functions and its mysterious connections with the classification of Lie groups. My "biological" list of seven elementary catastrophes thus found its place in a much wider framework. I did not understand its real nature till some years later (1970–72), when some people explained to me the theory of flat deformations of an analytic set. As you know, if we have an analytic set, i.e., a set defined by a set of analytic equations, and introduce a parameter deforming these equations, we get an analytic family. About these deformations of analytic sets, algebraic geometers define a very specific class, which they call *flat deformations*. Even now I am not sure I really understand the algebraic definition of flat deformation, which, in fact, arises naturally from the theory of flat modules in homological algebra. But, roughly speaking, a flat deformation is a deformation in which the fundamental class of the generic fiber remains of the same dimension when we move the parameters, i.e., we are not allowed to jump the dimension of the fundamental cycle of the complex analytic family when we perturb the parameters. It turns out that, if we are given the germ of an analytic set at one point, then we may try to find out only its flat deformations, and it can be proved by general theory that the family of all flat deformations of a given germ corresponds to an analytic space of universal flat deformations. This space itself can be given an analytic structure.

Staying at the IHES at that time, I was a colleague of Alex Grothendieck. But we were never able to talk or to have any kind of mathematical connection between ourselves. I tried sometimes, but, after a few minutes of talk, Grothendieck was immediately embarked in his own terminology, in his own way of describing things, and I was too lazy to follow Grothendieck's seminar and learn his terminology. The result was that we worked more or less independently of each other. Perhaps because of this, Grothendieck wrote me a personal letter saying that, in this period, I was too lazy. He might have been right in this. I was never able to do as he did: to work all through the night, to go to bed at 3:00 in the morning (or eventually later) typing his machine all the time. This is the type of work I am perfectly unable to do.

At that time, Grothendieck already had a general theorem about universal deformations of analytic sets; but this was, of course, a very abstract result, as was the work generally done by Grothendieck. It turns out, in this problem, that if we take an arbitrary singularity of an analytic set, then its universal deforming space is itself an analytic space, which is in general a very bad set to manipulate; it is in general of infinite dimension and has a large amount of singularities. If the singularity is isolated, then automatically the unfolding space becomes finite dimensional but with singularities. Trying myself to explain the origin of natural forms I had, of course, in mind the basic scheme explained in my first book *Structural Stability and Morphogenesis*, which I wrote in 1967–68.

In this scheme one starts by considering a natural medium, and one defines an equivalence relation between its points. Two points x and y in this domain are said to be equivalent if there exist local neighborhoods U of x and V of y, and a mapping g of U to V having the property that, if we pick a point z in U and take its image $z' = g(z)$ in V, then the local phenomenological properties of the medium in z and z' are the same. The neighborhood of x cannot be phenomenologically distinguished from the neighborhood of y. This defines an equivalence relation between points, and then one asks what are the equivalence classes.

This is a very natural thing to do for any kind of natural system. In some sense, the shape of any object is described by this sort of equivalence class of phenomenological equivalence between points. In geology, when people cut in a trench, they speak about the *facies* of the mineral and would put in the same stratum two points having the same facies. Two facies may be not exactly the same, but they may deform continuously one into the other. This is, in some sense, the basic scheme for descriptive morphology for any natural system. This was the original concept introduced in my book, originating what later became catastrophe theory.

This book was not immediately published. Initially entrusted to the American publisher Benjamin, the edition ran into difficulties, namely the publisher bankrupted and was taken up again by Addison–Wesley only in 1971–72. A few copies were however circulating under the hat and some found an enthusiastic reception by Christopher Zeeman at the University of Warwick. Zeeman helped a great deal in making it possible for the book to appear in French in the spring of 1972. His own reflections on this subject led to a grandiose extension of the theory. In *Stabilité Structurelle et Morphogenèse* I had restricted myself to a single universal substrate, namely $\mathbf{R}^3$ space, or, at most, space-time $\mathbf{R}^4$. But Zeeman, setting the theory ("catastrophe theory," as he called it) in the framework of a "general theory of systems," took for substrate any locally Euclidean space, sometimes defined in an abstract way by some semantic quality.

In fact, in general system theory one considers a system contained in a black box, having inputs occurring at discrete time (e.g. at integral values of time) and outputs at discrete values of time too. Suppose these are vectors. Then one gets a cloud of points in a vector space —product of the input vector space and the output vector space— and the basic program of general theory of systems is to deduce from the shape of the cloud of points the mechanism inside the black box, instead of smashing it into pieces and looking what is inside. The general system theory would look at the shape of the correspondence between inputs and outputs and try to find out, by a sort of interpretation of the shape of these *characteristics* (as they call them), what is inside the black box.

This way of looking at things was a radically new viewpoint opening the way to the creation of multitude of "models" that could be applied to the most varied disciplines, from physics (geometrical optics), to biology and human sciences. Invited to the Vancouver International Congress of Mathematicians in 1974, Christopher Zeeman delivered a dazzling lecture which he was later asked to repeat. As a result, catastrophe theory (which I will call CT from now on) took off like a rocket, propelled by the principal media all over the world. This glory was short-lived, however, and the brief success of CT soon fell to the slings and arrows of trans-Atlantic criticism.

Criticism and Defense of Catastrophe Theory

This experience played a central rôle in my subsequent "calling" as a philosopher of science. There were two main criticisms levelled against CT. The first went like this: Our universe is what it is, and if a phenomenon in it does not possess the "generic" form, there is no way of creating in the world the small deformation that could restore this generic form. Thus, for example, classical dynamics is described by Hamiltonian systems, which are not "generic." The second criticism bears on the pragmatic inefficiency of catastrophe models; the theorem of the unfolding of a singular point of a C^∞ function does not allow quantitative prediction, but at best a qualitative prediction in the neighborhood of the singular point.

It is not impossible to counter the first accusation to some extent; for example, one can invoke the implicit necessity of satisfying certain constraints. (In the case of Hamiltonian systems, a local micro-reversibility with energy conservation.) The second, on the other hand, seemed to me from the start to be fully justified. Since catastrophe models are defined only up to one C^∞ change of coordinates, we cannot use them for prediction in the way that we can apply quantitatively exact physical laws. I did not find it hard to admit this pragmatic shortcoming of CT (although Zeeman was more reluctant to admit it and, in fact, I am afraid he does not admit it even now). But, without subscribing to Rutherford's assertion "Qualitative is nothing but poor quantitative," I had to accept that an inefficient model might nevertheless be "worth thinking out," inasmuch as it confers on the global configuration of the system studied a global intelligibility which would be difficult to acquire otherwise —for example, through experimentation— or to describe by conceptual thinking linguistically expressed.

For those who know about the classical model of dogs' aggression which starts Zeeman's book *Catastrophe Theory*, it is true that I find this model still very useful in the sense that, if we want to express this model in a linguistic way, we cannot do it very easily. We have to use a lot of paraphrases all the time, and the thing which is very easily expressed by the mathematical picture is not easily described by ordinary language. I think this is the fundamental usefulness of catastrophe-theoretic models: They give a picture of a situation which is not immediately amenable to linguistic description.

Anyway, this controversy about the rôle and importance of CT, which finally ended in the years 1977–78 by a kind of disaffection of classical science with respect to CT, obliged me to think about what can be expected from science, and whether the general principle of science that everything has to be justified experimentally —this a priori dogma of experimental justification— has to be admitted.

Just at that time, the theory of catastrophes was superseded by the theory of the so-called *order to noise*, and essentially by the Prigoginian theory of dissipative structures. I was quite surprised by the fact that this theory won such a huge sociological success. In fact, Mr. Prigogine got the Nobel Prize for thermodynamic irreversible phenomena, of which practically nothing is known. I am still surprised that this was not obvious for the people in charge of the Nobel Committee that they could have crowned a nonexisting theory. Anyway, it is true that Prigogine forced most scientists to be aware of the importance of irreversibility in phenomena and about the irreversibility connected to morphological events in the medium. This is a great merit of the Prigoginian theory.

Nevertheless —getting back to my proper region of thought— the success of chaos was also, of course, a factor that took people away from the theory of catastrophes. CT was associated with local dynamics of potential functions, and potential functions is an extremely special type of dynamics. No doubt that it is a very special case, but, in some sense, it is perhaps the only dynamics for which the prediction problem can be solved. That is, given the initial data, can one predict where the trajectory will finish? This can be solved, in a universal way, only for dynamics made by gradients. Otherwise, in general, it is never known what is the outcome of a given trajectory. In that respect, I think that it is extremely important to understand first the simple case in which a complete prediction of the future of any trajectory can be given.

But the success of chaos started in 1975; essentially, after Ruelle–Takens theory in 1972 and, later, the theory of the so-called *weak turbulence*, which was also a great surprise among scientists, who did not expect turbulence to be deterministic. I must say that Arnol'd told me, already in 1966, that he was aware of the weakness of the Landau theory of turbulence and, in that respect, one should certainly recognize his position. But, very rapidly, people discovered the so-called *strange attractors*. Then came the classical examples of Lorentz, Rösler, etc.

From the point of view of practical applications for describing natural phenomena, I do not think —except for phenomena associated with physics— that we have a lot of examples of validity of the "chaosological" approach. Let us say that the idea, which was expected to be very useful, of explaining epileptic attacks by some sort of chaotic process, has turned out to be very "enervant" or quite inadequate.

Among physiologists, for example —there have been many physiologists interested in the chaos explanation of epilepsy, but there are already two radically different schools—, it is said that normal physiology is regular, and it is pathology which leads to chaos; whereas there are a few other physiologists which took the opposite position and said that the reverse is what is true: Normal state is chaotic and many pathologies appear by creating a very simple attractor. The typical example is, precisely, epilepsy. The normal

α-rhythm of EEG is extremely oscillating (there are a lot of nearby variations), whereas the epilectic feature is characterized by an extremely rigid, periodic attractor. So, this is a good example where interpretation of chaotic concepts in concrete phenomena leads to fairly difficult problems.

Going Back to Aristotelian Logic

Another criticism of CT was about analogy. Some people said that the bad thing about CT is that it leads to metaphoric thinking whereas the pure scientists do not admit analogy; they examine reality in itself. I think it is, more or less, philosophically an illusion to distinguish between reality and metaphor. In fact, analogy is, to some extent, a deep phenomenon of our thinking and if we want to understand what analogy is, then we are led to very fundamental philosophical problems, which have now been dropped, more or less, out of the consideration of people. But, in the Middle Ages, with Scholasticism (essentially, the ideas emanating from Aristotle), the problem of analogy was a fundamental problem, for the following reason. Aristotle defined what he called a *genus* as being a class of predicates, a class of adjectives, that had some sort of continuity property. Aristotle used this notion as a kind of classifying scheme for qualities or even for objects in the world. This led him to the notion that a genus might either be a subgenus of another genus, or it might be completely "uncommunicable." That is, it could not be added to other genera. In Platonic arithmetic, there were numbers that could not be added one to the other.

This kind of independence of objects is something very important if we think in the usual way. I was aware of this fact because, already in 1969, I wrote an article against the use of set theory in elementary schools, even in kindergarten. For instance, they put in front of a child a box containing some cubes; large and small, red and blue, and they asked the child to take out from the box the cubes that were *large or blue*. I do not know what the children did —maybe the most intelligent did it— but the fact is that, in usual language, the copula *or* cannot be placed between two adjectives which do not belong to the same genus. The opposition between large and small occurs in the genus of quantity, whereas the opposition between blue and red is an opposition in the genus of colour, and these genera are completely alien one to the other. Thus, asking the children to extract cubes that are *large or blue* is a task which is completely against the natural structure of the mind. We cannot say that a fellow is *short or intelligent*, because these two qualities do not belong to the same genus, whereas we can say that a fellow is *tall and intelligent*, or *short and intelligent*. When two genera are completely independent, we can put the copula *and* between the two qualities, because, in some sense, we take the transverse intersection of the two genera in the universal space of qualities. However, we cannot use *or*.

This is, of course, completely against the usual logic, for which the copulas *or* and *and* are considered as corresponding to union and intersection in the classical interpretation of first order logic. For me, first order logic is basically unsound from the point of view of usual thought. When Boole wrote his books saying that it was an investigation into the laws of thought, he was entirely mistaken. He was totally in the wrong direction from the way we really think, because this independence of genera of qualities is a very important structure of the mind which completely disappears if one takes for propositional logic the set-theoretic interpretation.

So this led me back to consider the old logic of Aristotle and to look at his metaphysics. As a result, I wrote a book which appeared recently in Spanish translation. I am happy to have the opportunity of doing some publicity, although this has been purely coincidental, because I received this translation just before leaving Paris. In this book, I tried to get back to the original ideas of Aristotle and discuss what kind of relation one can see between the way he looked at the world and my own view of applications of CT. There are basically two main points of concurrence between Aristotle and my point of view. First of all, it is obvious that we cannot use any kind of syllogism in Aristotelian logic. Take the following syllogism: "Any man is bipedal; a one-legged man is a man; hence, a one-legged man is bipedal." What is wrong with this reasoning? Of course, if you are a logically-inclined man, you would say that it is the first step which is wrong, because a one-legged man is not bipedal. But if you are a bit more philosophical, you would say "Well, it is natural to say that any man is bipedal, because normally a man is bipedal; there are very few exceptions, which are of course accidental, and we do not take account of these exceptions in the concept of man when we claim that all men are bipedal."

I think that Aristotle had the idea that logic has to be founded in the ontological nature of the concepts. It has to express something in reality. The fact that any man is bipedal is associated with a fundamental fact of biology, namely the fact that, expressed in modern terms, it is "inscribed" in the genoma of the animal that the animal normally developing will have two legs. This is associated in Aristotelian metaphysics with a well-known notion, which is what the old Aristotelian people of the Middle Ages called the *substantial form* of the concept. The concept has a sort of ontological ground in reality, which is its substantial form, its "essence." And it is part of the essence of the man to be bipedal.

So, if we want to use logic in a natural way, in a way which is founded in reality, we have to introduce this sort of essential quality which is associated with any concept, and we will have accidental references of this concept which modify this quality. If we take this point of view, the syllogism fails most times. In Aristotelian physics, the basic notion is that, in front of processes, one has to distinguish those which arise most of times and let pass those which are accidental.

This is a very sound way of looking at logic, and also a very sound way of looking at science itself. We should have a place for accidents for which we do not have any good explanations. In the point of view of CT, one makes the assumption of *genericity* and, when one wants to interpret morphology by means of dynamics, one always assumes the hypothesis of genericity. Doing this hypothesis one makes an appeal to the notion of "something occurring most of the time." Of course, genericity in the framework of smooth potential dynamics is a much stronger notion, because genericity implies an *open* property (the opposite case as, let us say, *meager* in the space of functions). Of course, this distinction did not occur to Aristotle, but the basic idea that natural processes are generic is, in some sense, fundamental in Aristotelian physics.

So I discovered that the things for which I took so much time to develop in CT were already known, to a large extent, by Aristotle. This is, essentially, what I express in my book *Esquisse d'une Sémiophysique*.

Qualitative Description Versus Exact Quantitative Modeling

Of course, it is not to deny that modern science was an advance. Modern science was an advance in the following sense. When Aristotle looked at the process of throwing a stone upwards vertically, he said the stone had first a sort of motion which he called a "forced motion." The upward motion for a stone is not a natural motion, because it goes against Nature. But, arriving at the top of the trajectory, the stone starts falling again, because it wants to reach the center of the Earth, which is the "natural locus" of the stone. So, for Aristotle, there is a sort of catastrophe at the top point of the trajectory, and then the motion of the stone changes its nature. You would probably say that now it is completely silly to look at this process in this way. First of all, I will tell you that saying that gravitation is described by the potential

$$V = gz$$

is perhaps no more explanatory than Aristotle's concept of natural locus. It is perhaps more precise, but it is not very much explanatory. The second point is that the great success of Galileo was to discover that the two motions had the same equation

$$z = v_0 t - \frac{1}{2}gt^2.$$

So, the falling motion was the analytic continuation of the ascending motion. Of course, this was a tremendous discovery, because once we know that some phenomena are directed in their evolution by a law associated with an analytic function, then we are able to make quantitative predictions. In fact, when we know part of a trajectory, analytic continuation gives —at least theoretically— the whole trajectory. We have a kind of canonical way of extrapolation and, hence, the possibility of prediction. I believe it is still true that, even now, strict quantitative prediction in science is associated with analytic continuation. As soon as we get out of the field of analyticity, analytic continuation is not possible. As a result, there is no strict way of extrapolation and no strict quantitative prediction becomes possible.

But the main point is that the qualitative description of a function is still useful in mathematics. In the first year at the university, if we want to teach people how to construct the graph of a function $y = f(x)$, we tell them first to compute the derivative, localize the zero points of the derivative, compute the value of the function at these points, draw the corresponding horizontal segments and then match these segments by a continuous curve monotonously increasing or decreasing between these segments.

This says that the qualitative description of a mathematical object remains useful in any case. This was the basic philosophy of Henri Poincaré. When, in 1880, it was found that the three-body problem was not solvable in any reasonable sense, Poincaré turned to qualitative study of differential equations in the plane and founded qualitative dynamics. This was a tremendous project which is, in some sense, a sort of anti-Galileo success. But now science is still very rigid; everything has to fit into the Galilean scheme. This concerns especially the people using computers, because for them everything has to be finally computed. Thus everything has to be described by, essentially, analytic processes.

But the number of phenomena which are amenable to specific analytic descriptions is relatively very small. We are not willing to accept this, because we think immediately about fundamental physics. Fundamental physics, of course, is expressed by laws which have an analytic nature. Nevertheless, these laws are analytic essentially because they follow from hypotheses of fundamental symmetry of the Universe. So, we have to make assumptions about the fundamental symmetry of the Universe to be able to derive laws explaining the motion of very huge phenomena, like, for example, the origin of the Universe out of the Big Bang and, finally, ending in the subquantum levels of very tiny particles.

People think that because one has laws for very large phenomena and also for extremely small phenomena, one should also have similar laws for phenomena in between. But this belief is quite likely unfounded. There are good reasons to believe that, in descriptions of theoretical physicists, phenomena like, for instance, metastability are still completely outside prediction. Take a glass pane in a window, which is a system in a metastable situation. According to the theory of statistical mechanics, after some time it should fall into its basic components: to a dust of glass particles. Nevertheless, we still believe that our glass will not break so easily, barring accidents.

So, a large number of natural phenomena in our scale still do not obey the general principle of exact quantitative prediction, and this should oblige us to take into account the possible data of qualitative approach. This is why I think that after some time people will, of course, say that CT is not able to give practical results, but nevertheless it may bring a lot of qualitative understanding that could not be easily obtained in any other way. It is my conviction that, after all this turn-over on CT and chaosology, one will get a most balanced appreciation of the situations.

Living as a Philosopher

All this is associated in this book with philosophical conceptions. As a result, I gained some reputation as a philosopher of science. I discovered, to my distress, that the sociology of philosophers is completely different from the sociology of mathematicians. Mathematicians form, in general, a community which —perhaps because of the Fields Medal after all— has a sort of worldwide unity. If we ask a generic mathematician to order the mathematical value of Mr. X with respect to Mr. Y or Mr. Z, in general we get a fairly substantial consensus among the corresponding ordering. When we look at philosophers, this is by no means the case.

I think philosophers live in their own nations; practically there is no worldwide community of philosophers, not even philosophers of science. In France, in particular, there are many philosophers who think that a good philosopher has to be a good writer, has to have style. Since Jean-Paul Sartre, we have been accustomed to think that a philosopher should write plays, novels, and so on. As a result, to do strict scientific philosophical work in France is something which is not very wealthy. I do not want to generalize too much, but certainly it is not something which is very important.

As I said, it is very difficult to know, even now, who are the most brilliant philosophers in Germany, in Italy, even in the United States. Of course, people in the United States have better opportunities of getting a reputation, essentially for editorial reasons still. This sort of situation of the philosophical public is something which is rather difficult to accept for the people who write outside their original speciality.

This is perhaps the lesson which has to be deserved among the Fields Medalists who might perhaps be tempted to give up their original speciality and specialize in other directions. They should not expect to find things easy. As I said, the community of mathematicians has a great knowledge of itself, yet, in the domains associated with human sciences, you would see that the situation is completely different.

I am particularly interested in linguistics. I think that my ideas have opened up some interesting aspects of linguistic theory but, up to now, although I have not been ignored by the linguists, nobody really cares about that. They say that these are things of no interest for them.

This is perhaps an aspect to be considered. I think we need to reach culture with respect to science, so that one finally recognizes that what is important in science is not the distinction between true and false. This might seem strange to mathematicians, but I will say that if I had the choice between an error which has an organizing power of reality (this could exist) and a truth which is isolated and meaningless in itself, I would choose the error and not the truth. There are many examples of errors which are scientifically important, and there are many, many examples of meaningless truths in science.

René Thom
Institut des Hautes Études Scientifiques
35 Route de Chartres
F-91440 Bures-sur-Yvette
France

Prepared from the author's text and the videotape of the talk, by Àngel Calsina.

Symposium on the Current State
and Prospects of Mathematics

Barcelona, June 1991

Rôle of Integrable Models
in the Development of Mathematics

by

Sergei Novikov

Fields Medal 1970

for his important advances in topology, the
best known being his proof of the topologi-
cal invariance of the Pontrjagin classes of
a differentiable manifold. His work included
a study of the cohomology and homotopy of
Thom spaces.

Abstract: The history of mathematics and theoretical physics shows that the starting ideas of the
best mathematical methods were discovered in the process of solving integrable models.

Mathematical discoveries of the last twenty years will be especially discussed as by-products of the
famous integrable systems of the soliton and quantum theories.

Rôle of Integrable Models in the Development of Mathematics

Starting as a Topologist

Before discussing certain models of mathematical physics and explaining their rôle in the development of modern mathematics, I want to say a few words about my own experiences as a mathematician. Let me start by reconstructing my career, not from the point when I was awarded the Fields Medal, but much earlier.

I started my mathematical life working in algebraic topology, and continued in this area for more than ten years; in fact, I still consider myself first as an algebraic topologist. When I started doing mathematics, in the mid-fifties, Russia was a very dark country, living behind the iron curtain. However, we had a very large and powerful mathematical school, whose leading person at that time was Andrei Kolmogorov in Moscow. He was the greatest mathematician, I think, after Poincaré, Hilbert and Hermann Weyl. A lot of famous mathematicians were his former students: Gel'fand, Arnol'd and many others (not me).

There was a common point of view in the Moscow mathematical school, concerning what was important and what was not important. The "important" areas of science were set theory, logic, functional analysis, and partial differential equations (not in the sense of solving models, but in the sense of proving rigorous theorems and establishing foundations). In Russian mathematics of that period —as in French mathematics— the main goal was to construct some kind of axiomatization, and the leading mathematicians were pursuing that. Topology was not existent in Russia in that period; there were only some remains of Pontrjagin's scientific school.

At the end of 1956, I was a second-level undergraduate student and had to choose one area —at least for some time— in order to be able to participate in seminars. I was attracted by an announcement posted in the Faculty of Mathematics and Mechanics of Moscow University. It was signed by Postnikov, Boltyanskiĭ, and Albert Shvarts. (The latter was a graduate student, but he was not considered as a "young mathematician;" in Russia, people aged 25 were not "young" in that period.) It was written in the announcement that there was a very new and exciting science, namely modern algebraic topology, opposite to the nonsense of point-set topology (maybe my translation is not very exact). These people were punished after that announcement; especially Shvarts (Postnikov and Boltyanskiĭ were professors, so life was somehow easier for them). Even Alexandrov, the famous topologist —who just continued a science which was thirty years old at that time— was terrified. So there was no place for Shvarts to continue his job in Moscow University, and he had to leave. Then he started learning about Fredholm operators; later, he moved to quantum physics and participated in the discovery of instantons, in cooperation with Polyakov. He was the first to discover nontrivial topological quantum field theory, ten years ago. In some sense, he followed the same way in science as me, but he was the most active during that period.

My friends —like Arnol'd and Sinaï, who were children of Kolmogorov's seminar, and Anosov, a child of Pontrjagin's seminar in control theory— asked me why I was trying to learn such a strange science, which was "completely useless," instead of studying important sciences like probability or partial differential equations. Thus topology was completely outside of the interest of our community in Russia. Postnikov told me that there were no prospects in topology, yet I could perhaps find something in cohomological algebra. Only Shvarts was enthusiastic about topology; however, he left Moscow very soon after he finished his thesis.

I published my first paper when I was 21. I was not "young" at that period because people like, for example, Arnol'd, wrote their first papers at age 18 or 19. This was completely normal. I come from a mathematical family, and my mother complained that "Everybody has published scientific papers, except my son."

I first worked in homotopy theory. Postnikov and Dyn'kin had made a very good translation of a collection of famous papers, mainly by French mathematicians: Serre, Cartan, Thom, ... We learnt them in our seminar. I was impressed by the excellent papers of the leading person in homotopy theory at that time: Frank Adams (who died recently). He started as an extremely brilliant scientist, solving famous problems.

This was a very interesting period. For example, nontrivial Hopf algebras —which are now very popular in the framework of quantum groups— were discovered during that time, shortly after the axiomatization of the work of Hopf by Armand Borel. The first persons who wrote papers about Hopf algebras were Adams and Milnor; before them, Hopf algebras were just cohomology rings of H-spaces and Lie groups. My first papers were dedicated to applications of Hopf algebras to the computation of homotopy groups of spheres and Thom complexes, which are important in cobordism theory. After that, I moved to the theory of differentiable manifolds, under the influence of several people who started visiting Russia at that period. John Milnor and Steve Smale helped us to start differential topology. Dynamical systems also started, in connection with topology and new kinds of algebra, after Milnor and Smale. When I write my memoirs, I will write something about all this, because I have found inaccuracies in many historical articles and books. For example, Smale's influence on the crucial point of the theory of dynamical systems in structural stability started in Russia. This relevant fact is missing in the historical literature, yet I was a direct witness of it.

Anosov and myself, together with Arnol'd, Sinaï, Shafarevich and Manin, organized a group of people who learnt different branches of mathematics from each other. Later Gel'fand's group joined us. People from partial differential equations started to interact with us after the discovery of the index of operators in the early sixties by Walter, a strange person from Bielorussia who appeared in Moscow at that time. This was done before the Atiyah–Singer paper; in fact, Atiyah and Singer wrote their paper after the publicity that Gel'fand made of Walter's achievements.

Topology started to be recognized as something serious more or less after 1961. The main question that I wanted to answer was "For what are we working?" As I said, I had good connections. I consulted friends like Arnol'd and Anosov; as a result, I got involved with foliations, in connection with problems of dynamical systems which were originated by Smale. Other friends helped me in connection with index problems from Gel'fand's school, by teaching me about partial differential equations —I also wrote something in that area— and learning topology in their turn. People working in algebraic geometry were also extremely useful; they helped me in the use of certain algebraic concepts in topology. However, I found out very soon that, no matter how far I was moving into mathematics, I was not able to answer my basic question, concerning the goal of what we were doing. I found that the theory of partial differential equations was as abstract as topology, and probability even more (I never worked in probability, but my friend Sinaĭ explained this to me; he moved from that area to dynamical systems himself). Dynamical systems was a much more beautiful and newer area; however, it played no rôle in the real world either, because nobody knew enough about it; it was too hard for people working in natural sciences in that period.

The Split between Mathematics and Physics

Arnol'd taught me, in his seminar, analytical mechanics and elements of hydrodynamics, in the framework of classical mechanics (not of quantum mechanics). Indeed, in the mid-twenties, after the creation of quantum mechanics, there was a very serious split between mathematics and physics in Russia (and not only in Russia). The best mathematicians of Kolmogorov's period —with a very few exceptions— never knew even the mathematical language of theoretical physics. The new language of theoretical physics started to be constructed more or less in 1925. According to physicists of that period, the crucial point in the divergence between mathematics and physics was not the creation of relativity —the rôle of relativity was realized later— but the creation of quantum theory. In Moscow, Gel'fand was probably the only one who learnt the new physics. Sometimes physicists participated in Gel'fand's seminar; however, in the late fifties Gel'fand stopped his job completely and physicists disappeared from that seminar for as long as twenty years.

I know a lot of childish tales from mathematicians about physicists, and from physicists about mathematicians, normally based on a lack of information. I have continually heard them for the last thirty years, even from great mathematicians or from the best physicists. I heard the latest yesterday: René Thom —one of my great teachers— spoke in his talk about the weakness of the "Landau theory of turbulence," which was pointed out to him by Arnol'd (who is also my friend, and with whom I have a lot of family connections). This was typically caused by a lack of information; one should not worry too much about these things happening. The story is that Landau never had any theory of turbulence. Landau had been interested in hydrodynamics since 1940, for twenty years at least, starting from his famous papers in superconductivity. For twenty-five years he was the only person who claimed that turbulence was purely a dynamical effect, a result of some global dynamics. It is a stupid idea, he said, to think that Navier–Stokes is nondeterministic: If one looks at a turbulent flow locally along time, one immediately observes that it is a very well defined flow; thus it is not reasonable to say that it is "something nondeterministic." Landau produced the idea that it is the result of something global, as a dynamical system. The deep point of his ideas was that nothing serious was known in the physics community about these new complicated examples of dynamical systems (even in pure mathematics they appeared relatively late). He said that these were perhaps some complicated infinite-dimensional tori embedded in a functional space. This viewpoint has been adopted by mathematicians, yet a considerable part of the ideology of Landau has been ignored because of lack of information. This may be the origin of the tale. (I should add that I have no interest in criticizing bad mathematicians. The criticism is only interesting if it is addressed to a good scientist. I would be happy if there is a revenge with the same weapons.)

Arnol'd, who is in some sense my teacher in questions of mechanics, started coming to my lectures in the early sixties, when I was 23 years old (I remember that he was one of the three people who attended my lectures). He was shocked by the idea of *transversality* and *generic position*. Transversality was a completely new concept, even for people who were famous in the theory of functions of real variables in 1961. Thus Arnol'd also learned something from me, while I learnt mechanics from him.

He told me that Kolmogorov proposed him to improve the result which is now called the *Kolmogorov-Arnol'd-Moser theorem*. Kolmogorov found the basic ideas and invited Arnol'd to continue them and furnish a rigorous proof. Kolmogorov also asked him to learn mechanics. Thus he read a lot of books, starting from Appell and some Russian books written by people in classical mechanics; however, as he said, he could not understand what mechanics really was. Then he found the book of Landau and Lifshits (which was not yet famous at that time among the mathematical community). He told me that, after reading this book, he finally understood what mechanics was, and, after that, he understood how bad the book was. Arnol'd himself wrote a brilliant book on mathematical understanding of classical mechanics. I would honestly say that I do not like that book, because he completely reconstructed the ideology. The book of Landau and his school was just a starting point to develop a great science; it contained many initial points allowing further progress. In Arnol'd's reconstruction, the mathematics is, of course, much better —it is a very good book for pure mathematicians—, but starting points for future research areas are missing. People who read Arnol'd's book arrive at an endpoint.

Learning Physics

My friend Manin had the same views as me about those books. We both independently decided, at the same time, to start learning quantum physics (my brother used to tell me that mathematicians should know everything about quantum theory). I first tried to learn quantum field theory as mathematicians normally do, and found out that this task was completely impossible. It might even be stupid to do so. Instead, I very much like the style of Einstein's lectures or the best lectures of Landau. I understood that *naturality* was the base of that science, exactly as I had earlier realized in topological books. The topologists of that period, like Jean-Pierre Serre, René Thom or John Milnor, sometimes omitted definitions in their lectures; they just said "This definition is natural." I recognized this style of "naturality" in the best physicists; in the lectures of Einstein, in the best books of Landau. (Not all books of Landau are equally good, but the collection of all of them is very valuable. Our students who want to do theoretical physics must know all these books at the age of 22. It is their common starting point.)

We discussed with Manin some paradoxes and unclear features of quantum theory, about which we shared a common point of view. I remember Manin telling me that every mathematician would find unclear points, but it would be a mistake to stop at those points and stand on criticism against them. Many mathematicians, including my students, have important difficulties in learning theoretical physics. They want to learn it as if it were mathematics: If they find something that they do not understand, they stop. I may definitely say that physicists also find a lot of nonunderstandable things; however, one must go ahead and think about such things only after having done a lot of exercises and reached a certain level.

It is very difficult to carry out Hilbert's program and to write theoretical physics in an axiomatic style. Hilbert did important work after Einstein's discovery of general relativity. He realized that the Einstein equation was an Euler–Lagrange equation for some functional. Thus he confirmed, in the case of the Einstein equation, that the axioms for any fundamental physics theory have to be started from a Lagrangian principle. Hilbert's program was useful for Hilbert himself, because he used it in that way. However, I do not like the experiences of some of my friends —extremely good mathematical physicists— who work in Hilbert's program, trying to make physics rigorous. This is, I think, impossible. One may prove a good theorem here, a good theorem there, about some physical situations. However, I think that Richard Feynman is completely right when he claims that it cannot be done globally (perhaps it is sometimes possible locally). The development of physics is more rapid than the flux of theorems which try to axiomatize it. The percent of things which may be done rigorously is going to zero; the number of good theorems is increasing, but the ratio is going down very rapidly.

I spent at least five years, between 1965 and 1970, just learning physics. Sometimes, half of my working time was dedicated to learning it as a student, from the earliest books of Landau, Lifshits, Einstein and others. In 1970, I wanted to make contact with physicists and people from Landau's school (they worked in the newborn Landau Institute, which was created in 1965). There were increasing rumours among physicists —even among engineers– that something very interesting was being developed, namely algebraic topology (of course, things like dynamical systems were also "topology" for them). Isaac Koletnikov, who was director of the Steklov Institute, informed me that people at Landau Institute had some very good problems in general relativity. They needed a topologist, so I joined them. We started working together at the end of 1970. At that time, I knew general relativity, which I had learnt earlier as a part of differential geometry. (But not in mathematical books; definitely, the best books on general relativity are not mathematical. It is better to read books by physicists, even in order to understand what is the best mathematics therein. Einstein's lectures are suitable to start with. Also the books of Landau–Lifshits, Misner, Thorne, are extremely good.)

Contributions to the Domain of Physics

The first paper which I was able to write was a joint paper with my collaborator Bogoyavlenskiĭ. He was my student at that time. We worked together in general relativity and just applied our knowledge about dynamical systems. My acquaintance with people in dynamical systems, which started ten years before, became fruitful in that period.

In this situation, I worked in the theory of homogeneous cosmological models, studying the space of homogeneous solutions of the Einstein equation, which leads to complicated dynamical systems, especially near cosmological singularities. There is an extremely unusual feature in this area, even from the point of view of the theory of dynamical systems. It is completely concrete; the ideas of genericity cannot be applied.

Physicists normally make the following criticism about such remarkable papers as the one by Ruelle and Takens. It is very good to construct abstractly nice examples of complicated dynamical systems. (A lot of such results were given before Ruelle and Takens, starting from Marston Morse in the thirties. Smale and our people —like Anosov, Arnol'd and Sinaĭ— also found a lot of remarkable examples of this kind.) But when dealing with real Navier–Stokes equations or with real Einstein equations, the question is "Is this situation realizable or not?" I may definitely say that it has never been realized up to now in hydrodynamics. Even remarkable specialists in ergodic theory, like Sinaĭ, were not able to understand, for ten years, how such Hamiltonian systems —like Einstein equations of homogeneous cosmological models— could lead to nontrivial ergodic properties, i.e., to nontrivial strange attractors. As far as I know, there is only one example of a strange attractor which can be investigated analytically. All concrete discoveries about attractors were only done on computers. In fact, mathematicians play a secondary rôle here; the mathematical job starts only after the picture is completely clear on the computer.

There is a particularly interesting geometry in the phase space describing the dynamical system of general relativity. In fact, some people from the Landau school had discovered these anomalous regimes before us; but nobody believed them, not even in the community of theoretical physicists. (I am not speaking about mathematics now; that job was totally done within the physics community.)

Bogoyavlenskiĭ and myself developed a certain technique of dynamical systems. As a special case of our technique, we were able to compare these strange anomalous regimes with strange attractors. They were computed analytically —not numerically— because of this very special strange geometry. But they were not *generic* attractors. In concrete systems, generic ideology does not work, because fundamental systems always have some important hidden symmetry.

We worked rigorously. This does not mean that we proved any theorem rigorously describing attractors. Rather, our skeleton was constructed in such a way that there were points with zero Lyapunov eigenvalues. We worked rigorously in the sense that our computations were done without arithmetical mistakes, in the process of solving those models.

At that time, my former student Bogoyavlenskiĭ presented his second dissertation in order to become full professor. It was the continuation of those methods on dynamical systems (later, he published a book in English about qualitative methods in gas dynamics and general relativity). The famous physicist, Zel'dovich, was the person who evaluated Bogoyavlenskiĭ's papers. He said that remarkable results had been obtained, not only in relativity, but also in gas dynamics. Moreover, after the official speech, Zel'dovich expressed to me unofficially how much he liked that technique, and how complicated it was to use it. I have to say that, at that time, even the elementary Poincaré two-dimensional plane qualitative theory was considered as a high-level theory in the physics community. Only the best people were able to use it. The level of nonlinear science in the community of the best theoretical physicists was very low.

Modern Developments in Topology

My own work at the Landau Institute has been divided into two parts. One part was dedicated mainly to topology (I was paid to work as a topologist, and people just consulted me about "modern mathematics"). A lot of topology appeared in this period, also related to instantons. I remember my friend Polyakov —who is six years younger than me— visiting me at home and asking "Did you hear anything about characteristic classes?" I told him that this was a very trivial concept from differential geometry. After I displayed the elementary formulae for dimension two (using tensor language; I could not use differential forms at that time), he said "It is quadratic!" Next day, he told me that he had found what has become the famous self-duality equation. Characteristic classes were quadratic, so it was possible to combine them with the Yang–Mills functional in that way!

I also interacted as a consultant with other people, leading to a famous discovery about the rôle of homotopy theory in liquid crystals and in anomalous kinds of superconductors in the early seventies. Fifty years before that, people had made interesting observations in optical experiments, like singularities in liquid crystals such as cholesterine (this is something that you can buy in a pharmacy and freeze up to -150 degrees in order to observe phases which display singularities). Nobody was able to realize before the early seventies that this phenomenon can be, in fact, explained by elementary homotopy theory. There were two groups working independently in that direction around 1974: one was in France; the other was at the Landau Institute.

I was not able to find anything substantially new and interesting in topology for myself until 1980. Before that, topology worked as it had been built. The new era began when topology started to apply discoveries of physicists inside topology itself. There was a huge noise in the physics community about topology during the seventies; also among applied physicists, working in low-temperature physics, liquid crystals, and related fields. It was a common opinion of physicists that the main new thing which physics borrowed from mathematics in the last ten years was topology. (You may find this assertion, for example, in an article written in the early 80's by Anderson, the famous applied physicist in superconductivity, and Nobel Prize.) A lot of new things indeed appeared during the eighties, as I will next explain.

Thus I started producing new work in topology in the past decade. I sold my knowledge about topology to physicists during that period, exactly as in my first years I had sold dynamical systems. In fact, dynamical systems was more my own way of solving models. The new direction in my work started with the discovery of soliton theory, which was done in the late sixties. It was an extremely interesting finding, which led, among other things, to modern integrable systems, conformal field theory, and quantum group theory (as a late by-product).

Integrable Models in Classical Mathematics and Mechanics

Everyone knows the rôle of the famous two-body problem, solved by Newton, in the development of the mathematical methods of physics. For a long period after that, people used the method of the exact analytic solution for some differential equations as a principal tool in mathematical physics. They simplified their problem if it was (or looked) too difficult, and after that tried to find the exact solution.

A lot of work has been done in the process of searching for special "integrable cases" of famous problems, like the motion of the top, for example. All mathematical methods —like power and trigonometric series, Fourier–Laplace (and other) integral transformations, complex analysis and symmetry arguments— were discovered and developed for that in the nineteenth century. These methods led sometimes to remarkable negative results, i.e., to proofs that certain models are not solvable in principle.

Some strange integrable cases which do not admit any obvious symmetry were discovered in the nineteenth century: the integrability of geodesics on 2-dimensional ellipsoids in Euclidean 3-space (Jacobi), the motion of the top with special parameters for constant gravity (Kovalevskaya), and some others. Riemann surfaces and θ-functions of genus 2 played the leading rôle in their integrability. What kind of hidden symmetry can be found behind this? It was not completely clear until the discoveries of soliton theory.

We have to add to these discoveries a new understanding of the deep hidden algebraic symmetry in the two-body problem, based on the so-called Laplace–Runge–Lentz vector of integrals. The "hidden" group generated by it acts on the energy levels in the phase space; this group is isomorphic to $SO(4)$ for the negative energy levels corresponding to the closed elliptic orbits, and isomorphic to $SO(3,1)$ for the positive levels corresponding to the hyperbolic noncompact orbits.

Generic spherically symmetric forces lead to the existence of nonperiodic orbits in an arbitrarily small neighborhood of any periodic orbit in the phase space (which is 6-dimensional for a particle in 3-space). The Kepler two-body problem is exceptional; all orbits are closed for any negative energy. As a consequence, any small spherically symmetric perturbation of the gravity forces leads to the famous displacement of the perihelium. This displacement was extremely important for the astronomical testing of celestial mechanics and general relativity. The group $SO(3)$ is not enough for the periodicity of orbits! A very large hidden symmetry is needed here.

We shall discuss later the fundamental rôle of the same symmetry for applications of quantum mechanics to the structure of atoms, discovered in the 1920's by Pauli.

Integrable Models in Quantum and Statistical Mechanics (1925-1965)

The integrable models of classical mechanics were forgotten by almost everyone except a few very narrow experts in these very classical problems. A new era of qualitative methods started with Poincaré. But the very new important branches of modern physics, like special and general relativity and quantum mechanics, were discovered in the first quarter of the twentieth century. Some initial problems for the Einstein and Schrödinger equations were also explicitly solved in the classical style.

For two particles with opposite charges, the nonrelativistic electric force is mathematically the same as the gravity force. The exact solution of this quantum nonrelativistic "two-body" problem preserves the same type of hidden symmetry as in classical mechanics; there is the quantized Laplace–Runge–Lentz vector commuting with the Hamiltonian (energy operator) and generating the Lie algebra of the group $SO(3,1)$ for positive energy and $SO(4)$ for negative energy. As a consequence, the spectrum of this quantum system (the Balmer spectrum) is more degenerate than the spherical symmetry requires (the group $SO(3)$). The stationary localized states have negative energy and can be considered as the quantized periodic orbits for the classical Kepler-type problem for electric forces. Together with the famous Pauli principle (two electrons cannot occupy the same state in any family of all possible states which present the orthogonal basis), this spectrum leads to an approximate explanation of the Mendeleev classification of elements.

One may observe that hidden symmetry of this sort is an exceptional property of the r^{-2}-type forces (and also of the linear forces) but these exceptional cases appear in the most fundamental problems of classical and quantum theory as a first good approximation. There is a common belief in the community of theoretical physicists that *the most fundamental mathematical principles of physics should be described (at least in some good first approximation) by objects containing enormously large hidden symmetry.*

It is very difficult to classify the wide collection of concrete problems solved by physicists in the process of developing quantum theory between 1925 and 1965. We want to point out the well known results of Bethe and Onsager, who solved some multiparticle one-dimensional quantum models and the models of statistical mechanics. The famous *Bethe Ansatz* for the construction of eigenfunctions was discovered and the two-dimensional Ising model was solved. But the influence of these discoveries on the mathematical methods of physics was fully recognized only much later, after the discovery of soliton theory in the process of quantization of its methods.

Soliton Theory

In the early sixties, people working in the theory of plasma observed that the Korteweg–de Vries (KdV) equation appears as the universal first approximation for the propagation of waves in many nonlinear media, combining nonlinearity and dispersion (if viscosity can be neglected). Before that, KdV was known for many years only as a very special system for waves in shallow water. People started to investigate the KdV equation in the sixties, and the most important discovery was made in the papers of Kruskal–Zabusky (1965), Gardner–Greene–Kruskal–Miura (1967), Lax (1968). This highly nontrivial system is in a sense exactly integrable by a very strange and new procedure (*inverse scattering transform*).

Let us be more precise now. The KdV system has a form

$$u_t = 6uu_x - u_{xxx}.$$

Its simplest solutions, known since the nineteenth century, are the *solitons*

$$u(x - vt) = -\frac{3a}{\text{ch}^2(3a)^{\frac{1}{2}}(x + 12at)}$$

and the *knoidal waves*

$$u(x - vt) = 2\wp(x - vt) + \text{constant}.$$

The soliton is localized and the knoidal wave is periodic in x. Here $\wp$ is the doubly-periodic Weierstrass elliptic function, whose degeneration is exactly the soliton.

Consider the Sturm–Liouville operator and the third-order operator

$$L = -\partial_x^2 + u(x, t), \qquad A = -3\partial_x^3 + 4u\partial_x + 2u_x.$$

Their commutator

$$[L, A] = 6uu_x - u_{xxx} = Q_1$$

is multiplication by the function Q_1. It means that the *Lax equation*

$$\frac{\partial L}{\partial t} = [L, A]$$

is formally equivalent to the nonlinear KdV equation.

Inverse Scattering Transform

The famous GGKM procedure may be immediately deduced from the Lax equation. Suppose all functions $u(x, t)$ are *rapidly decreasing* for $x \to \pm\infty$. Consider the special solutions of the linear Sturm–Liouville equation with exponential asymptotics

$$L\psi = \lambda\psi, \qquad L\varphi = \lambda\varphi, \qquad k^2 = \lambda,$$

$$\psi_\pm \to e^{\pm ikx}, \quad x \to -\infty,$$

$$\varphi_\pm \to e^{\pm ikx}, \quad x \to +\infty.$$

There is a unimodular transformation from the basis ψ to φ:

$$\varphi_+ = a\psi_+ + b\psi_-$$
$$\varphi_- = \bar{b}\psi_+ + \bar{a}\psi_-$$
$$|a|^2 - |b|^2 = 1.$$

The coordinates $a(k)$, $b(k)$ determine the so-called *scattering data* for the Schrödinger operator L with localized potential. The potential may be completely reconstructed by the *inverse scattering transform* from the function $[a(k), b(k)]$ with the proper analytical properties, plus a finite number of "discrete data," as was known long ago in the fifties.

The GGKM theorem states that:

(a) $\dfrac{da(k)}{dt} = 0, \quad \dfrac{db(k)}{dt} = (ik)^3 b(k).$

(b) The discrete eigenvalues are the integrals of motion.

(c) Local densities of the integrals of motion for KdV may be constructed in the following way. Consider the formal solution for the Riccati equation

$$\chi_x + \chi^2 = u - \lambda, \qquad \chi = k + \sum_{n \geq 1} \frac{P_n(u, u_x, \dots)}{(2k)^n}.$$

Then the integrals

$$I_m = \int P_{2m+3}\, dx$$

are the local conservative quantities for KdV.

(d) Exact (multisoliton) solutions will be obtained from the *reflexionless potentials* $b(k) \equiv 0$.

KdV hierarchy. Hamiltonian properties. Generalization

There is an infinite number of operators

$$A_0 = \partial_x, \quad A_1 = A, \quad \dots \quad A_n = \partial_x^{2n+1} + \cdots, \quad \dots$$

such that the commutator $[L, A_n]$ is multiplication by a certain polynomial

$$Q_n(u, u_x, u_{xx}, \dots, u_{2n+1}).$$

The *KdV hierarchy* is the collection of nonlinear systems

$$\frac{\partial u}{\partial t_n} = Q_n,$$

equivalent to the collection of Lax-type equations

$$\frac{\partial L}{\partial t_n} = [L, A_n].$$

All these flows commute with each other; we may find a common solution

$$u(x, t_1, t_2, t_3, \dots), \qquad t_0 = x.$$

The detailed study of the polynomials Q_n was done in the late sixties by Gardner. In particular they have the following form (*Gardner form*)

$$Q_n = \frac{\partial}{\partial x}\left(\frac{\delta I_{n+1}}{\delta u(x)}\right)$$

$$I_{-1} = \int u\, dx, \qquad I_0 = \int u^2\, dx, \qquad I_1 = \int \left(\frac{u_x^2}{2} + u^3\right) dx, \quad \dots$$

As observed by Gardner, Zakharov and Faddeev (1971), this is the form of the Hamiltonian equation corresponding to the GZF-Poisson bracket

$$\{u(x), u(y)\} = \delta'(x - y)$$

and to the Hamiltonian

$$H_n = I_{n+1}.$$

The inverse scattering transform may be treated as a functional analog of the transformation from $u(x)$ to the action-angle variables as in analytical mechanics, which were useful for the semiclassical quantization (Zakharov and Faddeev).

The quantization program was started after 1975 by Faddeev, Takhtadzhyan, Sklyanin and others in Leningrad. Different groups starting from 1971 discovered many new interesting nonlinear systems integrable by the Lax equation and inverse scattering transform. Some famous systems —well known before— were solved by that method (for example, nonlinear Schrödinger, sine-Gordon, discrete Volterra system or discrete KdV, Toda lattice and many others, including some special two-dimensional systems which proved to be very important later).

Starting from 1976, different groups found other interesting properties and generalizations of the Hamiltonian formalism of KdV theory. Interesting new Poisson brackets with large hidden algebraic symmetry were discovered; for example, the Lenart–Magri second bracket for KdV, and the Gel'fand–Dikiĭ brackets for the generalizations of KdV hierarchy to the scalar operators of higher order.

The Periodic and Quasiperiodic Problems. Riemann Surfaces and θ-Functions.

The appropriate approach to solving the periodic problem for KdV was found by the present author (1974). After that, it was completely solved by Novikov and Dubrovin (1974), Lax (1975), It·s and Matveev (1975), McKean and Van Moerbeke (1975) and by Krichever (1976) for the (2+1)-system of KP-type (Kadomtsev–Petviashvili). The basis of this approach is the KdV-invariant class of "finite-gap" Schrödinger operators with periodic and quasiperiodic potentials, whose spectrum on the line **R** has a finite number of gaps only. These potentials satisfy the stationary KdV and higher KdV equations

$$\left[L, \sum_{j=0}^{N} c_j A_j \right] = 0,$$

where N is the number of finite gaps.

This system is a completely integrable finite-dimensional Hamiltonian system. Its solution was found explicitly in terms of the θ-functions of a Riemann surface whose genus is equal to N:

$$u(x,t) = \text{constant} - 2\partial_x^2 \log \theta(Ux + Vy + U_0),$$

where U, V are certain N-vectors.

Finite-gap potentials form a dense family in the space of continous periodic functions (Marchenko–Ostrovskiĭ, 1977) and even in the space of quasiperiodic functions. For the periodic potential $u(x + T) = u(x)$ the Schrödinger operator L commutes with the shift $\hat{T} : x \mapsto x + T$. Therefore, there is a common eigenfunction for the operators L and $\hat{T}$ (which is familiar in solid state physics)

$$L\Psi = \lambda\Psi$$
$$\hat{T}\Psi = e^{ipT}\Psi.$$

Here $p = p(\lambda)$ is some multivalued function of λ and the spectrum is exactly the set of all λ such that $p(\lambda) \in \mathbf{R}$.

There are exactly two Bloch eigenfunctions $\Psi_{\pm}$ for each complex value of λ (except for a countable number of branching points). The function Ψ is meromorphic on some hyperelliptic Riemann surface Γ (two-covering of the λ-plane), generically of infinite genus. For the real potential $u(x)$ on the line $\mathbf{R}$ the branching points are real; they exactly coincide with the endpoints of the spectrum. This is the *spectral curve* Γ.

The generic real nonsingular solution of the commutativity equation

$$\left[L, \sum_{j=0}^{N} c_j A_j\right] = 0$$

is quasiperiodic (it contains a dense family of periodic solutions with different periods). All of them have the spectral curve Γ of finite genus N and vice-versa. The potential $u(x)$ and the eigenfunction Ψ may be expressed in terms of θ-functions corresponding to the surface Γ. There was a very interesting formal algebraic investigation of the commuting ordinary differential operators in the twenties (Burchnall and Chaundy); a Riemann surface was discovered based on the polynomial relation between L and A, which they found for any commuting pair, with no formal connection between periodicity and our surface. There is a theorem stating that, for operators of relatively prime orders with periodic coefficients, these two surfaces in fact coincide. It may be not so in more complicated cases.

The general elementary idea between the appearance of Riemann surfaces in the theory of finite-dimensional integrable systems can now be explained. It is very useful to relate the original Lax representation for KdV-type systems to the compatibility condition of the two linear 2×2 systems whose coefficients depend on λ (*zero curvature equation*):

(a) Lax equation (1968)

$$\frac{\partial L}{\partial t} = [L, A].$$

(b) Zero curvature equation (1974, for KdV hierarchy and sine-Gordon)

$$\frac{\partial \Psi}{\partial t} = \Lambda(\lambda)\Psi$$

$$\frac{\partial \Psi}{\partial x} = Q(\lambda)\Psi$$

$$\frac{\partial \Lambda}{\partial x} - \frac{\partial Q}{\partial t} = [\Lambda, Q].$$

For the ordinary KdV we have:

$$Q = \begin{pmatrix} 0 & 1 \\ u - \lambda & 0 \end{pmatrix}$$

$$\Lambda_1 = \Lambda = \begin{pmatrix} -u_x & 2u + 4\lambda \\ -4\lambda^2 + 2\lambda u - u_{xx} + 2u^2 & u_x \end{pmatrix}.$$

For higher KdV we have the same Q; the corresponding matrices Λ_n are polynomial in λ. For the stationary equation ($\partial_t = 0$) we have "Lax-type" representations for the finite-dimensional system

$$\frac{\partial \Lambda}{\partial x} = [Q, \Lambda].$$

The Riemann surface Γ appears here. It has genus one for the ordinary stationary KdV, and the Weierstrass elliptic function appears:

$$\det(\Lambda - \mu I) = P(\lambda, \mu) = 0.$$

The function $\psi(x, P)$ is meromorphic on the surface Γ. Its coefficients are the integrals of the system (they do not depend on x).

For KdV and higher KdV, all matrices Λ_n have zero trace and the Riemann surfaces are hyperelliptic:

$$\mu^2 = R_{2n+1}(\lambda),$$

where n is the genus of Γ.

The function ψ was explicitly found from the "algebro-geometrical data": The Riemann surface Γ and the poles of Ψ (Ψ has exactly n poles, which do not depend on x after the proper normalization). This approach works for all known nontrivial integrable systems. As is now known, this mechanism is valid also for the classical systems mentioned above (Jacobi, Clebsch, Kovalevskaya, ...). An important discovery was made by Krichever (1976), who improved this approach and found the generalizations to the $(2 + 1)$-systems like KP. *All Riemann surfaces appear in the theory of KP.* This last property is very important for applications of these ideas to different problems (like Schottky-type problems in the theory of θ-functions, analogs of Fourier–Laurent bases on Riemann surfaces —which are useful in string theory—, the formalism of τ-functions, etc). *The classical functional constructions were greatly extended after the periodic theory of solitons.*

Conclusion

To conclude this long discussion, we now present the collection of the different branches of mathematics and theoretical physics which were involved in the integrable soliton models. Important new connections between them were discovered.

(1) Nonlinear waves in continuum media (including plasma and nonlinear optics).

(2) Quantum theory; scattering theory and periodic crystals.

(3) Hamiltonian dynamics.

(4) Algebraic geometry of Riemann surfaces and Abelian varieties (θ-functions).

Quantization of the soliton methods led to the discovery of the *quantum inverse transform*, started by Faddeev, Sklyanin, Takhtadzhyan, Zamolodchikov, Belavin and others. An important new object, the so-called *Yang–Baxter equation*, started to play the leading rôle. Later (Sklyanin, Drinfel'd, Jimbo) Hopf algebras appeared here with the special *universal R-matrix* of Yang–Baxter, which led to the now very popular *quantum groups*.

Recall that from the self-dual Yang–Mills equation (Polyakov and others) the theory of instantons appeared, with remarkable applications in four-dimensional topology (Donaldson and other people in Atiyah's school). The theory of Yang–Baxter equations led to the Jones polynomials in the theory of knots. The famous 2-dimensional conformal field theories also gave us the remarkable collection of integrable models which are also the most beautiful new algebraic objects. They have deep connections with soliton theory and quantum groups.

As a result, I may formulate the following thesis: *A large part of the most important discoveries in mathematics and in the mathematical methods of physics was done in the process of developing the theory of integrable models.*

Sergei Novikov
Landau Institute for Theoretical Physics
Academy of Sciences of the USSR
GSP-1   117940 Kosygina Str. 2
Moscow V-334
Russia

Prepared from the author's text and the videotape of the talk, by Regina Martínez.

The Current State and Prospects of Geometry and Nonlinear Differential Equations

by

Shing-Tung Yau

Fields Medal 1982

*for his contributions in differential equations,
also to the Calabi conjecture in algebraic geometry, to the positive mass conjecture of
general relativity theory, and to real and complex Monge–Ampère equations.*

Abstract: In the development of mathematics in the twentieth century, multidimensional manifolds
have been one of the central subjects of research. In the last ten years, theoretical physicists have needed
some results on infinite-dimensional manifolds, which appear in modeling certain problems; for example,
a system with an infinite number of particles in statistical mechanics. Some of these results could be
translated to finite-dimensional manifolds. Perhaps some day even old classical problems can be solved
by applying some of these infinite-dimensional theories directly to universal moduli space.

A second area of important future development is the study of singularities, which occur practically
in every field. For instance, the algebraic theory of singularities has been developed in the last forty
years, but the application of this theory to natural systems related to partial differential equations will
be more than welcome.

Finally, modern computers generate many interesting questions. It is conceivable that deep mathematics may arise from understanding computer calculations in the future.

The Current State and Prospects of Geometry and Nonlinear Differential Equations

It is an honor to be invited to give a talk at this symposium.[1] It is certainly a difficult or even impossible task to talk about the future of mathematics. Mathematics is no longer an isolated branch of science. The interaction of mathematics with other subjects in both pure and applied science is increasingly vigorous. It is difficult to imagine that the discussion of the development of mathematics will not involve serious discussions of other branches of science.

On the other hand, how can we dare to say that we have enough feelings about other subjects? It is already difficult to find a mathematician who can claim to be an expert in many different branches of mathematics. Therefore, I will attempt to discuss subjects that I am more familiar with. For example, I am not capable of commenting on problems related to algebra, number theory or logic, etc. I will attempt to comment on developments related to geometry and analysis.

Within pure mathematics, nonlinear analysis interacts with topology, algebraic geometry, number theory, group theory and operator algebra. On the other hand, nonlinear analysis and geometry have become fundamental tools to such diverse subjects as particle physics, general relativity, inverse problems in quantum chemistry, modeling in biology, weather prediction, information theory, computer graphics and robotics.

In the following, I will discuss current developments and speculations for the future. Our observations and speculations are clearly based on personal taste. Although mathematics is both an art and a science, the solution and the selection of our problems is very much bounded by the development of historical processes. It is true that mathematicians can create problems without caring about their applicability; the motivation usually comes from attempting to understand some known phenomena. Such phenomena can either come from nature or from some other mathematician's work. A theorem is often judged to be important either if it can be explained elegantly, or if it can be applied to explain some natural phenomena from different areas, including physics or engineering. I will therefore follow such criteria.

---

[1] I was not able to attend the symposium due to the critical illness of my mother. She passed away before the conference. I am deeply disturbed by her death. Since my father passed away twenty-seven years ago, she devoted her whole life to educating me. I dedicate this paper to her memory.

Analysis on Infinite-Dimensional Manifolds

In the development of mathematics in the twentieth century, functions of several variables or multidimensional manifolds have been the central subject for research. Research on these subjects has called for unification of many disciplines in many subjects. Gradually, it is being recognized that there are natural infinite-dimensional manifolds.

A classical and important example is given by the problem of finding periodic solutions in Hamiltonian mechanics and Riemannian geometry. Such a theory goes back to Poincaré, Birkhoff, Morse and others. It is most successful for finding closed geodesics on a compact manifold. By studying critical points of the length function on an infinite-dimensional manifold, Morse, Bott, Klingenberg, Gromoll–Meyer, Sullivan and others were able to demonstrate that most compact simply-connected manifolds admit infinitely many closed geodesics. The homotopy type of the free loop space is deeply involved. Equivariant Morse theory and Morse theory with degenerate critical points were developed. In the past fifteen years, the *mountain pass lemma* —which can be considered as a version of the critical point theory— was successfully applied by Rabinowitz, Nirenberg and others in studying closed orbits for Hamiltonian mechanics, existence theorems for semilinear elliptic equations or existence of periodic solutions for hyperbolic equations.

When studying the corresponding questions for Hamiltonian mechanics, critical point theory for symplectic geometry was naturally introduced. There was the beautiful conjecture of Arnol'd on the possible number of fixed points for symplectic maps in relation to critical points of functions. Conley and Zehnder developed a beautiful theory to study critical point theory for symplectic manifolds. Floer was able to apply it to prove Arnol'd's conjecture for a large class of symplectic manifolds. Floer also applied the closely related intersection theory of Lagrangian submanifolds to construct Floer homology. The ideas of Witten on the analytic interpretation of Morse theory were also used by Floer.

In the sixties, Palais and Smale formulated the right condition for Morse theory to hold on an infinite-dimensional manifold. Unfortunately, in many important applications, we do not expect the full Morse theory to hold, except in the case of the loop space. However, partial Morse theory does hold in many concrete situations and it often involves new ideas. Morse–Tomkins and Courant already knew that certain parts of Morse theory work for minimal surfaces. The deep work of Morrey and Sacks–Uhlenbeck certainly gives a good understanding for this partial Morse theory. Similar phenomena occur in the later important works of Taubes and Uhlenbeck on gauge theory in dimension 4 and Gromov's work on pseudo-holomorphic curves in symplectic geometry. These theories basically tell us how the "missing" critical points disappear. It would be nice to develop a condition similar to the Palais–Smale condition to capture this kind of phenomena. Hopefully we would have a generalized Morse theory where the space of critical points can be "compactified" and the contribution of ideal critical points can be calculated.

It seems that, in the future, certain infinite-dimensional manifolds will be studied in their own right. This may not be too surprising, as systems with an infinite number of interacting particles are certainly interesting in themselves. In fact, in the last ten years, theoretical physicists have demanded that many theorems which hold for finite-dimensional manifolds should also be true for infinite-dimensional manifolds. Some of these formal analogies have implications for finite-dimensional manifolds. It is amazing that a lot of these implications are actually true and can be proved indirectly. Famous examples are given by the work of Taubes on defining the Casson invariant and the proof of Witten's conjecture on the rigidity of the elliptic genera. Clearly, certain parts of analysis and geometry can be generalized to infinite-dimensional manifolds in a rigorous manner. Can the spectral theory of the Dirac operator over the loop space be understood? It is clear that many of these problems need new ideas. This should lead to many new interesting questions for finite-dimensional manifolds. Topological quantum field theory, matrix model and many aspects of conformal field theory have already led to the introduction of new invariants for three- and four-dimensional manifolds, new conjectural identities for Chern numbers for moduli spaces and new questions in algebraic geometry. After Donaldson introduced his invariants through classical means, Witten was able to interpret them in terms of topological quantum field theory. The interpretations based on field theory are brilliant. However, they do not lead to proofs in the classical sense. Perhaps some day even old classical problems will be solved rigorously by applying some of these infinite-dimensional theories.

In the past twenty years, theoretical physicists, in their attempt to unify different forces, have introduced many important concepts including supergeometry and supersymmetry. Several infinite-dimensional algebras or Lie groups, such as the Virasoro algebra, the Kac–Moody algebra and quantum groups were introduced. These infinite-dimensional symmetries are necessary because we are dealing with infinite-dimensional objects. The representation theory of these algebras have broad and deep consequences in nonlinear differential equations and geometry. They are playing fundamental rôles in understanding various completely integrable systems in both fluid mechanics and statistical mechanics. The concepts of supergeometry and supersymmetry are still mysterious from the point of view of classical geometry. However, they lead to many beautiful developments. For example, they lead to a new proof of the Atiyah–Singer index theorem. They also lead naturally to a rich theory of compact algebraic manifolds with trivial canonical bundle. The symmetries provided by conformal field theory have provided a totally unexpected "mirror symmetry" in the moduli space of these manifolds. In particular, they provide a conjectural formula for counting the number of rational curves on the quintics threefold. The outcome is part of a very active research.

Modern theoretical physics has provided a very effective leadership for many activities in geometry. It is mysterious that such a theory can provide deep intuition in mathematics and yet such intuitions rarely lead to a direct rigorous proof. Perhaps a much better understanding of infinite-dimensional manifolds may help to resolve some of these mysteries.

Most of the mathematically interesting quantum field theories fix the underlying topology of the manifold. For example, in studying the Chern–Simons Lagrangian over the space of connections on a fixed vector bundle, the Donaldson polynomial was found. However, in quantum gravity we are supposed to allow the topology of the underlying manifold to change as well. Yet no significant mathematically interesting questions have been generated from these aspects of quantum gravity which allow the topology to change. In the future, if a unified field theory exists, we should see a mathematically interesting theory of geometry and analysis allowing the topology to change.

It is generally agreed that in the theory of strong interaction and gravity, nonperturbative effects are important. On the other hand, these effects are difficult to estimate. Extensive numerical calculations are needed. Since the dimension goes to infinity, a straightforward numerical calculation is not possible. The desire to match theory with experiments should drive theorists to pay more attention to computing. On the other hand, without a good theory it is difficult to tell if a numerical calculation is reliable. Hopefully, numerical computation for systems with an infinite degree of freedom will be developed to the extent that it can be used to predict physical phenomena reliably. Analysis on infinite-dimensional manifolds should be crucial for such a development.

Singularities

The concept of singularity arises in practically every field. Usually, when the structure is not smooth or "nongeneric," singularities exist. Singularities are not necessarily bad. In fact, they provide valuable information about the global structure. The best example is provided by Morse theory. However, we are still far from having a complete theory of singularities in geometry. Most fruitful activities are centered around geometric objects that are embedded in a nonsingular space. A closed set in Euclidean space defined by a set of polynomials naturally exhibits singularities at those points where a neighborhood cannot be parametrized nicely. The fundamental theorem of Hironaka says that, in this case, the singularity can always be resolved. In other words, by abstractly replacing the singularity with a suitable set of lower-dimensional subvarieties, the closed set becomes nonsingular. This theorem is clearly fundamental and has numerous applications. However, its relation to differential geometry has not been very fruitful yet, partially because the lower-dimensional subvariety provided by Hironaka is not too "canonical."

For a long time, there was no good cohomology theory or homotopy theory to describe singular algebraic manifolds. Ten years ago, Goresky and MacPherson introduced their concept of intersection cohomology. There is much evidence that the intersection cohomology is the right cohomology theory for singular spaces. The solution of the Zucker conjecture by Looijenga and Saper–Stern is certainly an important indication. Saper also constructed complete Kähler metrics on the complement of the singular set so that the L^2-cohomology is the same as the intersection cohomology. Stern has developed an L^2-index theorem for locally Hermitian symmetric spaces. For a large class of singular spaces, Tian and the author have constructed the canonical complete Kähler–Einstein metric. Hopefully, these metrics will enjoy good properties, so that for both the L^2-cohomology and the L^2-index, they will behave like locally symmetric spaces.

Besides complete Kähler metrics, there is also a class of Kähler metrics which reflect the geometric nature of the singularities. These are the induced metrics from the complex projective space. Not too many deep properties are known for this class of metrics. Since they are quite natural, we should expect more research in understanding the spectral theory of the metrics. For example, what is the L^2-index for the natural operators for these metrics?

When one considers the moduli space of various structures in algebraic geometry, one naturally obtains singular spaces and these are usually endowed with a natural metric which we call the *Weil–Peterson metric*. The geometry of this metric should be very interesting and the singularities of this metric should reflect the property of degeneration of the structures.

Another class of singularities occurs in the calculus of variations. In order to prove existence in partial differential equations, we enlarge the class of functions, so that compactness theorems can be used. The major work is to prove that the singularity set of these generalized solutions is either empty or small. A very important example is provided by the study of minimal submanifolds.

The theory of singularities for minimal subvarieties is very deep and has a long history. When the subvariety is two-dimensional, we understand the singularities quite well. Although there is more work to be done, we know roughly how topology changes when we minimize the area of a surface. There are many questions for the topology of three-dimensional manifolds that can be reduced to the question of how the topology of minimal surfaces changes when they are formed in a certain manner. There are much deeper questions related to unstable minimal surfaces.

For higher-dimensional minimal subvarieties, the problem is far more complicated. The best one can say about the set of singularities is that it is a closed set with certain codimension. Almgren has proved the best theorem, stating that the set is closed with measure zero. It is an important and difficult problem to study the structure of the set of singularities. Can the set of singularities be triangulated? A very similar problem occurs for harmonic maps between manifolds. The best work concerning singularities of harmonic maps is due to Schoen–Uhlenbeck. However, this work is not strong enough to prove any existence theorem. This is the case because we do not know how topology changes when we minimize the energy of the map.

In proving the existence theorem for critical points in the calculus of variations, we often study the gradient flow. As a result, we often study parabolic equations, even if we are only interested in elliptic equations. If a certain part of Morse theory holds, then parabolic equations often give more information. However, singularities occur more frequently for nonlinear parabolic equations. In fact, singularities create serious questions about the meaning of the evolution equation after the singularity is encountered.

Scalar parabolic equations are better understood. However, for vector-valued evolution equations, the situation is far more complicated. If the nonlinearity occurs only in the lowest order term, then there is an extensive literature and one can say much more about the solution. However, most evolution equations that occur in nature and in geometry contain higher nonlinear terms. A lot of these evolutions exhibit both parabolic and hyperbolic nature. Theoretically, the basic question is whether singularity can develop when the initial data is nonsingular and asymptotically well-behaved. One of the famous problems in mathematics is whether singularity can be found in the Navier–Stokes equation or not. The question is not yet known for the Euler equation, where most people believe that singularity should form.

Amongst evolution equations that do develop singularity, equations for gas dynamics with space dimension one are probably the best understood. Glimm, Lax, Liu, DiPerna and others have made deep contributions. However, when the space dimension is greater than one, practically no theoretical work is known.

In the last ten years, the nonlinear Schrödinger equation has been extensively studied. Papanicov and others have carried out both theoretical and numerical studies of the singularities of this equation. It can be a useful model for other nonlinear evolution equations.

There are many nonlinear hyperbolic equations in geometry. Most of them have not been studied. Even in space dimension one, the problem tends to be very difficult because the characteristics are curves that are difficult to understand. This already occurs in the problem of isometric embedding of surfaces with negative curvature in $\mathbf{R}^3$. When the curvature changes its sign, the equation becomes of mixed type. The structure of the singularities is so complicated that one is unable even to speculate what may happen.

The best known evolution equation in geometry is the study of the evolution of a surface governed by its mean curvature. When the space dimension is one, or when the initial data is a convex surface in $\mathbf{R}^n$, singularity does not develop, as is shown by Grayson, Huisken and others. However, in other cases, singularity does develop. It is of interest to know how it develops. At present, there is no well-accepted theory of generalized solution of this evolution equation. Osher and Sethian have studied this equation numerically from the point of view of flame propagations. Evans and Spruck studied the equation theoretically by replacing the surface by the level set of some function in higher-dimensional Euclidean space. It is possible that the definition of solution after singularity develops depends on the physical or geometric problem that we are modeling. In any case, this is a major problem to be studied.

Another important evolution equation was introduced by Hamilton. This is the first parabolic equation on deforming a Riemannian metric, which had a great success in the past ten years. There will be active research on this equation. The major question is how singularity develops. It is clearly related to topological changes of the manifold. Besides the potential applications for settling questions in topology, it may give a better understanding of constructing canonical metrics on compact manifolds.

A very subtle problem of singularity occurs in general relativity. A serious problem arises because both space and time can change topology. An apparent singularity can be "cured" by changing the coordinate system as Kruskal did in the case of the Schwarzschild solution. Besides the question of investigating singularity when both time and space are finite, it is also important to understand the singularity at infinity because it will capture the dynamics of the space-time. Unfortunately, no known explicit solution of the Einstein equation has ever been written which exhibits dynamics. This makes it very difficult to ascertain what is true, even partially, for gravitational radiation. On the other hand, there is the famous conjecture of Penrose called *the cosmic censorship*. It says that singularities of a generic space-time cannot be naked. Such a conjecture is important both philosophically and mathematically. It is very deep and very far from being solved.

There was some attempt in the past to analytically continue the Lorentzian metric to a Riemannian metric. It was a miracle that, for many classical solutions with singularity, the associated Riemannian metric is nonsingular. Perhaps this is a good way to understand singularity in general relativity.

At present, the most important question in classical relativity is to provide a concrete example of space-time with gravitational radiation. The structure of singularities at both finite and infinite distance are equally important. It will also be interesting to find a ten-dimensional nonsingular vacuum solution compatible with string theory.

Finally, we should mention the beautiful picture of singularities created by iteration of nonlinear maps on $\mathbf{R}^2$. The subject is being pursued by many distinguished mathematicians. We hope that these pictures will reveal some singularities occurring in nature. Apparently it is being successfully applied to pattern recognition. The ability to draw a nice picture is certainly important for geometry. However, only partial understanding of these pictures has been accomplished. Much more systematic research is still needed.

Classification of Geometric Structures

It has always been a central problem in geometry to construct and classify certain geometric structures over a fixed smooth manifold. In most cases, we are interested in when we can cover manifolds by coordinate charts where coordinate transformations are given by certain pseudo-groups or certain Lie groups. The most familiar pseudo-group is the pseudo-group of holomorphic transformations and the two most popular Lie groups are the group of conformal transformations and the group of projective transformations. Because of the potential applications to other fields, we will see the study of many other pseudo-groups or Lie groups. For example, in order to understand real algebraic manifolds, the pseudo-group of algebraic transformations will play an important rôle (the degree of algebraic transformations should be essential in estimating geometric quantities of the manifold).

An immediate consequence of the existence of a pseudo-group structure or a Lie group structure is the reduction of the structure group of the tangent bundle to a certain subgroup. Whether the structure group can be reduced to a subgroup is a much easier question than the existence of a pseudo-group structure. For example, it is not difficult to check whether a manifold admits an almost complex structure. But it is much more difficult to find a complex structure on a given manifold.

A basic question is whether every almost complex manifold with dimension ≥ 3 admits an integrable complex structure or not. There are two other important classical questions to be answered: How do we decide whether a complex structure admits a Kähler structure? Can every Kähler manifold be deformed into an algebraic manifold?

The fundamental work of Mori enables algebraic geometers to generalize the classification theory of algebraic surfaces to higher-dimensional varieties. However, many fundamental questions on algebraic surfaces remain to be answered. One of the basic questions is which smooth four-dimensional manifolds admit an algebraic structure. The description of the moduli space of the algebraic structure over these manifolds is also a fundamental question. The beautiful work of Donaldson made use of the topology of moduli of stable bundles to obtain uniqueness theorems in topology. Notice that Donaldson's work depends on the fundamental existence theorem of Taubes on self-dual connections.

Existence theorems are very important in geometry. One way to prove existence theorems is through partial differential equations or calculus of variations. Hodge theory is such an example. Another approach is to use algebraic analysis. Mori's proof of the existence of rational curves is such an example. It is very important to combine these approaches. An important example of such a unification is the development of the Riemann–Roch formula through Hirzebruch, Grothendieck and Atiyah–Singer.

For a large class of algebraic manifolds, the author is able to attach a canonical Kähler–Einstein metric. Any geometric invariants constructed from the metric will then be invariants of the algebraic structure. Interesting examples are provided by the spectrum of the metric. The works of Donaldson and Uhlenbeck–Yau also provide canonical Hermitian metrics on stable bundles. One can also study the metric invariants created by these metrics, which provide ways to characterize quotients of Hermitian symmetric domains. It is very desirable to see an algebraic proof of these uniformization theorems, because it should give a better understanding of the algebraic fundamental group of algebraic manifolds.

When the manifold has a pseudo-group structure such that it has a submanifold left invariant by the pseudo-group, we can discuss structures induced on the submanifold from the ambient manifold with a given pseudo-group structure. Compact submanifolds of this type are important in understanding the structure of the ambient manifold. Important examples of this type are algebraic subvarieties of a given algebraic manifold. It is always highly desirable to understand whether a homology class can be represented by this class of submanifolds. It is also important to count and to investigate how they intersect. The famous Hodge conjecture is an attempt to characterize homology classes that can be represented by algebraic cycles. The most recent development of "mirror symmetry" on the space of algebraic threefolds with trivial canonical bundle provides a way to count rational curves. It depends on the existence of supersymmetry which in turn depends on the existence of a Kähler metric with zero Ricci curvature.

In looking for submanifolds with structures induced from the ambient manifold, one is naturally led to look for bundles with certain structures. Sections with certain properties then give the cycles. A fundamental question is the existence of certain structures on a given topological bundle. If the structure is holomorphic, a necessary condition is that all the Chern classes are of (p, p) type. If a topological bundle has this property, then one expects that, after adding a holomorphic bundle, the total space will admit a holomorphic structure. We can consider this statement as a generalized Hodge conjecture.

When the structure is modeled after a noncompact Lie group, it is very much related to the theory on noncompact Lie group actions on a manifold. Such a theory is also related to the study of bundles with a connection whose holonomy group is a subgroup of a noncompact Lie group. Both theories are being actively pursued. Margulis' famous work forms the crux of most modern developments. His powerful methods have been the basic tools for the last fifteen years in the theory of discrete groups acting on manifolds with nonpositive curvature.

Thurston's uniformization for three-dimensional manifolds is a theorem on the existence of $SO(n, 1)$-structures on a manifold. This is the best existence theorem on G-structures with given topological data. Either weakening the hypotheses or generalizing to higher dimensions would be very important. One should take Thurston's approach as an important guideline for the future development of existence theorems with topological data.

Many years ago, the author initiated the use of harmonic maps to study the rigidity of discrete group actions and the complex structure of a manifold. This line of research was successfully followed by Siu, Sampson, Eells, Corlette, Gromov, Jost, Schoen, Carlson–Toledo and others. The other line of research comes from the study of Yang–Mills connections. Donaldson, Uhlenbeck–Yau and Simpson made contributions to these considerations. We hope to see more analytic approaches to discrete group theory.

Finally, we should mention that one of the most fascinating structures in geometry is the existence of Einstein metrics. There is no general approach besides the constructions of special classes of homogeneous Einstein metrics or Kähler–Einstein metrics. This is a very unsatisfactory state of affairs, because we are unable even to speculate as to how to approach the problem. This is because we know very little about geometric properties of Einstein metrics, especially when the scalar curvature is negative.

There is an extensive literature on the space of metrics whose curvature is pinched between two constants. We hope that these may eventually lead to insight into existence theorems.

Classical Differential Geometry

It is a pity that the beautiful subject of surfaces in $\mathbf{R}^3$ has not been pursued vigorously in the past twenty years. The Russian geometers led by Alexandrov, Pogorelov and Effimov made many contributions to classical questions in geometry. The intensive development of computational geometry and mechanical engineering will clearly stimulate much more activity in this classical subject.

Bending of surfaces in $\mathbf{R}^3$ provides a lot of attractive but difficult questions. The most outstanding is the possibility of deforming closed surfaces isometrically. Donnelly has given a piecewise-linear example without boundary. However, no smooth example has ever been found.

Isometric immersion of surfaces encounters great difficulties when the curvature of the surface is negative or changes its sign. This is partially due to our ignorance of nonlinear hyperbolic equations. In the case of isometric immersion, it is due to ignorance of the behaviour of asymptotic curves. The most outstanding achievement is Effimov's generalization of Hilbert's theorem that surfaces with strongly negative curvature cannot be complete. It is still difficult to describe their singularity set; it is conjectured that singularities cannot be isolated.

Local isometric immersion is not well understood, although C. S. Lin has proved beautiful theorems on isometric embedding of non-negatively-curved surfaces.

Russian geometers have developed a very complete theory for convex surfaces, and it would be very desirable to generalize most of their theory to nonconvex surfaces. Much fruitful research has been done on the surfaces obtained by calculus of variations. This is especially true for minimal surfaces, capillary surfaces and surfaces with constant mean curvature. For the latter part, Wente has solved the outstanding Hopf conjecture. We shall see many more developments in these directions.

Higher-dimensional submanifolds are much more difficult to understand. Despite the famous *Nash isometric embedding theorem*, we still do not know how to isometrically embed an abstract manifold in a "nice" manner. Much research needs to be done in this direction.

There has always been a strong desire to represent an abstractly defined structure by embedding it into some classically defined simple space. In the case of the Nash embedding theorem, it is not only a beautiful, but also a practical theorem. For example, in order to define weakly differentiable harmonic maps between manifolds, Morrey isometrically embedded the image manifold into a Hilbert space, because the Nash embedding theorem was not available at that time (Morrey's idea has independent interest because it was later used by Gromov to define geometric invariants). What is lacking in the Nash theorem is the control of the extrinsic quantities in relation to the intrinsic quantities. The "rigidity" is also far from being understood.

A very similar situation occurs in algebraic geometry where Kodaira's embedding theorem provides an embedding into CP^n. However, the defining equations are not well understood even for many simple intrinsically defined algebraic manifolds. In the case of complex manifolds or Kähler manifolds, we do not even have a candidate for them to embed into. How do we recognize them?

Shing-Tung Yau
Department of Mathematics
Harvard University
Science Center 325
One Oxford Street
Cambridge, Massachusetts 02138
USA

Symposium on the Current State
and Prospects of Mathematics

Barcelona, June 1991

Noncommutative Geometry

by

Alain Connes

Fields Medal 1982

*for his contribution to the theory of opera-
tor algebras, particularly the general classification and a structure theorem for factors of
type III, classification of automorphisms of
the hyperfinite factor, classification of injec-
tive factors, and applications of the theory of
C^*-algebras to foliations and differential geometry in general.*

Abstract: Through algebraic geometry we became familiar with the correspondence between geometrical spaces and commutative algebra. The aim of this talk is to show an analogous correspondence, in
the domain of real analysis, between geometrical spaces and algebras of functional analysis, going beyond
the commutative case. This theory is based on three essential points:

1. The existence of many examples of spaces which arise naturally, such as Penrose's space of uni-
verses, the space of leaves of a foliation, the space of irreducible representations of a discrete group,
for which the classical tools of analysis lose their pertinence, but which correspond in a very natural
fashion to a noncommutative algebra.

2. The possibility of reformulating the classical tools of analysis such as measure, topology and
calculus in algebraic and Hilbertian terms, so that their framework becomes noncommutative, the
commutative case being neither isolated nor closed in the general theory.

3. The relationship with physics, the spaces used by physicists being noncommutative in many cases.

Noncommutative Geometry

I would like to give a general survey of noncommutative geometry. I will explain the motivation and the general program. For this, I will rely mainly on two things. First, the oldest example in noncommutative geometry, which goes back to the discovery of quantum mechanics by Heisenberg; I will then continue with pure mathematics, and will end by coming back to physics (in fact, by coming back to what may be extracted from the actual phenomenology of elementary particles about the fine structure of space-time).

So let me begin by explaining what the general motivation for noncommutative geometry is. There is a well-known duality which occurs, for instance, in algebra and geometry between a space and a commutative algebra. Given a space, we want to study it by looking at the algebra of coordinates on the space, which has to satisfy a certain regularity. Of course, if we are doing algebraic geometry, we restrict ourselves to polynomial or algebraic functions; but when dealing with topology or differential geometry the regularity is less restrictive, and, for instance, we use continuous functions or smooth functions.

The basic theme of noncommutative geometry is that there are several quite important cases in which one is forced to replace a commutative algebra of coordinates with a noncommutative algebra.

The first instance of this goes back to Heisenberg, and the second example is the need to consider spaces or manifolds which are not simply connected and whose fundamental group fails to be Abelian (arbitrary finitely presented discrete groups may occur in this way). For these spaces, the ordinary use of the Pontrjagin dual of the group is, of course, inefficient. The third example comes from foliations: If the space of solutions of a differential equation is treated as a classical space, then most of the standard tools completely lose their pertinence; in fact, this space is precisely an example of a *quantum space*, in the sense that it is described by an algebra of coordinates that fails to be Abelian. Finally, the fourth example, which is now quite fashionable, is *quantum groups*. Let me explain in one word what quantum groups are in this framework: Quantum groups are the analog of Lie groups in noncommutative geometry.

A brief sketch of the general program is as follows. We wish to be able to transplant into the noncommutative setting the tools that we are accustomed to in the classical commutative framework.

When looking at a space in the classical way, there are a number of points of view which are like "finer and finer," and enable us to comprehend the space. The coarsest of these points of view is measure theory. If we know the space only up to measure theory, then essentially we know nothing, because most spaces are isomorphic in measure theory: They are isomorphic to the unit interval with the Lebesgue measure. Then we have topology and differential geometry (by "differential geometry" I mean only the differential geometry of differential forms, currents, characteristic classes, i.e., excluding the differential geometry which comes from a Riemannian metric). The fourth and most important point of view is thus Riemannian geometry. I will sketch at the end what the relevance of the noncommutative analog of Riemannian geometry is for the physics of elementary particles.

First Example: Quantum Mechanics

I will start now by explaining the origin of the subject, which is the discovery by Heisenberg of quantum mechanics. I would like to show how much this discovery relied on experiments and got rid of the usual framework of classical mechanics, forced by the experimental results. Let us go back to Heisenberg, at the time when he discovered quantum mechanics (which was not called "quantum mechanics," but *matrix mechanics* for a reason which will become clear when we look at the way it was found).

At that time, by a great deal of work, people had already realized that the atom was formed by an inner nucleus, around which there were revolving electrons that governed the chemical properties of the atom. Moreover, a fairly good way of observing atoms was by interaction with electromagnetic radiation. For instance, if one takes a prism and allows sunlight to pass through it, then this light will be decomposed into various rays, and, of course, these rays will form the colors of the rainbow. However, if one takes pure bodies like helium or hydrogen and looks at their emission spectrum, then this emission spectrum will not contain all the rays in the sunlight. It will only contain certain rays, which essentially form a sort of "signature" of the elements in question. Thus, it is extremely important to be able to understand the regularity of these rays. Now, if we try to apply classical mechanics in order to understand this, then we take for the atom a very simple model. Using mathematical language, this model will be described by the so-called *phase space*, which is known to be a symplectic manifold, and the functions on this space will be the *observable quantities* of the system. I have been thinking of the system as being the atom with the electrons and the nucleus, and all the observable quantities evolving with time according to the *Hamiltonian evolution*, which is given by the following equation:

$$\dot{f} = \{H, f\}, \tag{1}$$

where $\dot{f}$ is the time derivative of f, and $\{H, f\}$ is the Poisson bracket between f and a certain observable which is called the *energy*, which is the Hamiltonian of the system.

Now, for simple systems like the hydrogen atom, this equation will be totally integrable, which means that there are invariant tori describing the motion of the system, and in these tori the motion is almost periodic. This tells us that each observable quantity can be computed and expanded as a function of time

$$q(t) = \sum q_{n_1,\dots,n_k} \exp(2\pi i \langle n, \nu \rangle t), \tag{2}$$

where the coefficients $q_{n_1,...,n_k}$ are complex numbers, and $\langle n, \nu \rangle = \sum n_j \nu_j$ is a combination of the basic frequencies with the same integers n_j that appear in $q_{n_1,...,n_k}$.

If we take this mechanical system that describes the atom and try to describe in classical terms its interaction with radiation, then the answer is given by the Maxwell theory. Maxwell theory tells us that, when the atom is in interaction with radiation, it emits plane waves and these plane waves can be completely described as follows. Take the observable quantity, which is called the *dipole moment*. (What we have are electrons which are revolving around the nucleus; these electrons have a certain charge, and so they form a dipole moment around the nucleus.) This defines a certain observable quantity $\vec{Q}(t)$ which has three components and can be expanded in an almost periodic series

$$\vec{Q}(t) = \sum \vec{q}_n \exp(2\pi i \langle n, \nu \rangle t). \tag{3}$$

Maxwell theory tells us that any of the components $\vec{q}_n$ provides a plane wave W_n which has frequency $\langle n, \nu \rangle$. Thus, in particular, the observable frequencies should form a subgroup of the real line, generated by the basic frequencies ν_j.

It turned out, however, that observation was already giving at that time a result which was contradictory to this fact. If one observes, for instance, the spectral rays of the hydrogen atom, then one finds that the wavelengths of these rays are certain precise numbers which are, as I said before, a sort of signature of hydrogen. The regularity of these rays was already found by Balmer long ago. He observed that the wavelengths of these rays were all simple rational multiples of a certain length L. They are of the form

$$H_\alpha = \frac{9}{5} L \,, \quad H_\beta = \frac{16}{12} L \,, \quad H_\gamma = \frac{25}{21} L \,, \quad H_\delta = \frac{36}{32} L \,, \quad \ldots \tag{4}$$

so what we are dealing with is really

$$\lambda = \frac{n^2}{n^2 - 4} L. \tag{5}$$

The first thing that people realized then was that it was much more natural not to talk about wavelengths (it is the wavelengths one observes when looking at the spectral rays), but to talk about frequencies, which are calculated as the speed of light divided by wavelengths. When we look at frequencies, we get a simpler formula

$$\frac{1}{\lambda} = \frac{R}{m^2} - \frac{R}{n^2} \,, \tag{6}$$

where R is a constant called the *Rydberg constant*, and m and n are integers. Now, from the experimental results we find that the observed frequencies do not form a group; that is, they do not form a subgroup of the real line. What happens, however, is that if we look at them (see figure 1) we see that they combine together. For instance, we can take the first ray in the *Lyman series* (1-2) and combine it with the first Balmer ray (2-3), obtaining the second Lyman ray (1-3). If we combine it instead with the second Balmer ray, then we get the third Lyman ray, and so on. So what happens is that they do not form a group, but they combine according to the so-called *Ritz-Rydberg combination principle*, which is the following: One can label the frequencies by two indices, say ν_{ij} (these two

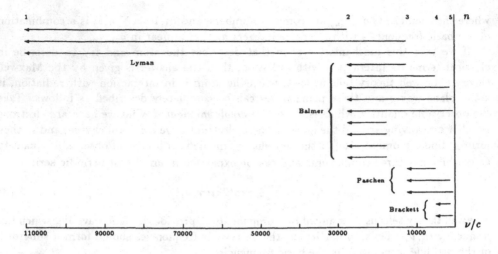

Figure 1

indices have nothing to do with integers; they may be whatever we want: Greek letters, colors, ...), and they combine according to the rule

$$\nu_{ij} + \nu_{jk} = \nu_{ik} . \tag{7}$$

This is what was found experimentally. Heisenberg used the following extremely pragmatic kind of reasoning: If one does a little bit of mathematics, then one finds that, in the classical case, there is an alternative way of describing the algebra of observable quantities. One takes almost periodic functions, which have the given frequencies, and multiplies them together by forming the convolution product

$$(qq')_{n''} = \sum_{n''=n+n'} q_n q_{n'} . \tag{8}$$

What is obtained is nothing other than the algebra of convolution of the group Γ, which is supposed to be the group of observable frequencies. However, as experiment shows, these do not form a group, so Γ is to be replaced by the set

$$\Delta = \{ \, \nu_{ij} = \nu_i - \nu_j \, \} \subset \mathbf{R} \tag{9}$$

of real numbers combining according to the Ritz–Rydberg combination principle. Heisenberg decided to follow the experimental results and to replace the commutative convolution algebra of the group Γ by the convolution algebra of the set Δ, and therefore to work out the convolution algebra of this set with its partially defined composition. It was found that the product of two observable quantities a and b was given by

$$(a \cdot b)_{ik} = \sum_j a_{ij} b_{jk} . \tag{10}$$

It is remarkable that Heisenberg invented this rule from experimental results, although he did not know about matrices. Later, he talked to Bohr and Jordan and found out that these things existed in mathematics and were called "matrices." This is why the theory was called *matrix mechanics*.

Then the law of evolution is quite simple; namely, it is given by

$$q_{ij}(t) = q_{ij} \exp(2\pi i \nu_{ij} t), \tag{11}$$

and something occurs which is even simpler than in the commutative case. Remember that the evolution was described by the Poisson bracket (the Poisson bracket was an additional structure which was coming from the symplectic structure of the phase space). Now, in the noncommutative case of matrices, this is not needed. It is replaced by the commutator

$$\frac{d}{dt} q(t) = \frac{i}{\pi} [H, q], \tag{12}$$

where H is a matrix which is zero outside the diagonal and whose diagonal entries are ν_i's such that $\nu_i - \nu_j = \nu_{ij}$. (This value ν_i is not unique; it is unique only up to a common addition of a constant.)

As a consequence of this, from the experimental results we cannot stick to a classical phase space; that is, we cannot stick to classical mechanics. Instead, we are obliged to replace the commutative algebra of functions of observable quantities of ordinary phase space with a noncommutative algebra.

It turns out that, in the case of Heisenberg, if we look at a system having finitely many levels (or even countably many) then the algebra that we get is not very complicated to analyse. But, for instance, as soon as we handle situations like quantum statistical mechanics —where one takes an assembly of atoms— then the noncommutative algebra that we are dealing with is much more difficult to analyse. But this would only pertain to the "measure theory" part of the discussion.

The Novikov Conjecture

After this motivating example of Heisenberg, I would like to enter the domain of pure mathematics, and deal with an example in which noncommutative geometry may be seen at work. The example is the following. We meet noncommutative objects as soon as we try to handle manifolds which are not simply connected. In fact, when we take a manifold M, to this manifold corresponds a group $\Gamma = \pi_1(M)$, its fundamental group, which measures the non-simply-connectedness of the manifold. Many of the results which are true for simply connected manifolds require more work when one tries to adapt them to the non-simply-connected world. Roughly speaking, the idea is that when one wants to adapt them to the non-simply-connected situation, one is no longer going to handle, for instance, vector spaces over the complex numbers, but modules over the group ring of a group. Basically, one is always taking into account the equivariance with respect to the action of the fundamental group, and, instead of doing things down in the manifold, one essentially has to work in the covering space. So I will try to show how noncommutative geometry works in examples, by dealing with the problem of the signature of a manifold.

I will first describe the signature theorem of Hirzebruch. If we take a manifold of dimension $4k$ which is compact and orientable, then there is an intersection form on the middle-dimensional cohomology and, by construction, the signature of this quadratic form turns out to be homotopy invariant, because it is defined in a homotopy invariant way. It is not clear at all whether it is possible to relate this quantity to other quantities which are computed, for instance, by characteristic classes of the tangent bundle of the manifold.

The following result is due to Hirzebruch:

$$\text{Sign}(M) = \langle L(M), [M] \rangle; \tag{13}$$

that is, the signature can be computed by pairing the fundamental class $[M]$ of the manifold M with a universal polynomial $L(M) = P(p_1, \ldots, p_k)$ on the Pontrjagin classes of the tangent bundle of the manifold, which depends on the dimension of M.

There is a huge difference between the two sides of the equation (13), since the left-hand side is homotopy invariant by construction, and the right-hand side is essentially computable by local computations (by integration over the manifold). This is a fairly good answer for the simply connected case, and Novikov proved that this specific combination of characteristic classes is the only one that can be homotopy invariant.

Things get more interesting when the manifold is not simply connected. In the non-simply-connected case, there are quantities which are candidates for being homotopy invariant, but for which it is not obvious at all, at first sight, that they will be. These quantities are the *Novikov higher signatures*, which are defined as

$$\text{Sign}_c(M) = \langle L(M) \cdot \varphi^*(c), [M] \rangle. \tag{14}$$

That is, we keep the same L-genus, but we have to be careful about one thing: Now the L-genus is not homogeneous, but it has several components. It has one component which is the top-dimensional component, and it also has components whose dimension differs from the dimension of M by multiples of 4. We multiply it by a group cocycle c of the fundamental group, after transferring it to a cohomology class on the manifold. (Since the manifold has a fundamental group Γ, the group cohomology of Γ maps very naturally to the cohomology of the manifold.) Then we compute the product $L(M) \cdot \varphi^*(c)$ and evaluate it on the fundamental class of the manifold M.

This is a fairly algebraic expression. For a more geometric definition, imagine that the so-called *classifying space* of the group Γ has been constructed. This is a certain space $B\Gamma$ that can be explicitly given in many cases. Essentially, we are taking a cocycle in $B\Gamma$, transversely oriented, considering a classifying map $\varphi \colon M \to B\Gamma$ transverse to the cocycle, and taking the inverse image $\varphi^*(c)$ of the cocycle.

The question is whether or not this signature is homotopy invariant. This is a purely geometric question, known as the *Novikov conjecture*. Novikov conjectured that in several cases, these quantities, which are called *higher signatures*, are homotopy invariant.

I would like to show how noncommutative geometry works in this case. Let me begin with the commutative case. I will first specifically discuss the situation when the fundamental group of the manifold is commutative, and we shall see how to make use of this commutativity. This is the proof given by Lusztig in this case. We shall see how many more tools we have when this group is commutative than in the noncommutative case.

If the fundamental group Γ is commutative, then it has a Pontrjagin dual $X = \hat{\Gamma}$, which is a compact space: the space of all linear characters of the group; that is, all homomorphisms from the group to the complex numbers of modulus one.

Now we can consider a product space $M \times X$ of the manifold and the Pontrjagin dual of its fundamental group. On this space we have a very canonical line bundle which is given by the fact that whenever we have a character, this character gives us a map from the fundamental group of M to the complex numbers of modulus one, and therefore a completely natural flat bundle on the manifold M with holonomy given by the character. Thus we get a family of flat bundles on M parametrized by the Pontrjagin dual X; that is, a natural line bundle on the product $M \times X$. It is not difficult now to consider the signature of this family (or the family of signature operators: For each of these flat bundles we have a signature with coefficients in the flat bundle, so that we can consider this family of operators). On one hand, the signature family is not just the difference between the dimension of the positive eigenvectors and the dimension of the negative eigenvectors; it is the subspace of the positive eigenvectors minus the subspace of the negative eigenvectors. What we really have is two vector bundles over the base X, and what we get in this way is not just a number, but an element of the so-called K-theory of X, which is denoted by $K(X)$. Once we have this element in the K-theory of the space X, it is not difficult to show, firstly, that this element is homotopy invariant (this is not much harder' to prove than that the ordinary signature is homotopy invariant), and, secondly, that if one takes the Chern character of this family —just applying the Atiyah–Singer index theorem for families— one gets exactly the Novikov higher signature.

Now, the problem which I really would like to deal with is what replaces the Pontrjagin dual X, the K-theory of X, the Chern character, the index theorem, and so forth, when the group Γ (the fundamental group of the manifold) is no longer commutative.

So far, we have used the commutativity in an essential way. It was used in order to define the Pontrjagin dual, and to deal with this Pontrjagin dual in the standard way of commutative spaces.

The Group K_0

What is K-theory? K-theory is essentially doing linear algebra with parameters that vary continuously in a base space X. If we view things algebraically, K-theory is just doing linear algebra where the ground ring has been replaced from the complex numbers $\mathbf{C}$ to the ring $C(X)$ of continuous functions on X. There is a purely algebraic definition of K-theory in terms of classification of finite projective modules. The fact that modules are finite corresponds to the fact that fibres are finite-dimensional, and *projective* is a translation of the fact that bundles are locally trivial. So, in fact, the meaning of doing K-theory over an algebra can be formulated in a purely algebraic way. Moreover, we find out very quickly that the commutativity of the algebra —the ground ring we are dealing with— has nothing to do with the problem. Therefore, we are free from the hypothesis of commutativity of $C(X)$. As soon as we are dealing with finite projective modules, we need to represent them inside matrices over the algebra, as idempotents. But matrices over an algebra do not commute, so commutativity has nothing to do with the problem.

The second point is that if we take a discrete group Γ, then the construction of the Pontrjagin dual gives a noncommutative C^*-algebra rather than a commutative one. Let me explain how this is constructed. One takes the regular representation of the group in the space $\ell^2(\Gamma)$, the Hilbert space with orthonormal base formed by the elements of the group. In this Hilbert space, the group is acting by the left regular representation and so the group ring —the linearization $C\Gamma$— is also acting. We simply take the norm closure of this group ring. If the group were Abelian, what we would get would be precisely the continuous functions on the Pontrjagin dual of the space (this is not difficult; it is just Fourier analysis). So, in general, we have a good replacement for this, except that it is not a commutative algebra.

There is a natural way to define the signature of the covering space of a manifold. If we look at the universal cover of the manifold, then on this universal cover we have the fundamental group acting, and we can mimic the usual construction of the signature on the universal cover. We can still consider the differential forms with a certain growth at infinity, and the cup product, which gives us a pairing and hence a quadratic form. It turns out that this quadratic form can be defined as an element of the so-called *Witt group*. The Witt group is a group of abstract quadratic forms over the group ring $C\Gamma$. But the trouble with the Witt group is that it is defined abstractly by a presentation of quadratic forms (we want them to be equal if they differ by a change of variables, or if they are stably equal, and so forth). So it is difficult to analyse.

Now it should be clear why we take C^*-algebras. Precisely because C^*-algebras are the only algebras for which the spectrum of the self-adjoint elements is real. Why not take for example the algebra $\ell^1(\Gamma)$ of summable functions on Γ? This is a Banach algebra. But if we take a self-adjoint element in it, in general its spectrum will fill in the whole corona. It is not true for an involutive algebra in general that the spectrum of a self-adjoint element is real; this is precisely the characterization of C^*-algebras. Therefore, we take C^*-algebras precisely in order to be able to say that an element of the Witt group —a self-adjoint quadratic form $H = H^*$ that belongs to the ring of $q \times q$ matrices over an algebra $\mathcal{A}$— determines a positive eigenspace and a negative eigenspace. How do we get these positive and negative eigenspaces? When the spectrum is real, we do a Cauchy integral over a closed curve C enclosing the positive spectrum of H

$$\frac{1}{2\pi i} \int_C R_\lambda \, d\lambda, \tag{15}$$

where R_λ is the resolvent of the quadratic form. In doing this, by general results, we know that we get an idempotent projection. So this enables us to say that the Witt group in this situation maps to the K-theory (and, in fact, the Witt group is equal to the K-theory). Thus for C^*-algebras the main simplification is that the K-theory is the same as the Witt group, and K-theory is far simpler since it is just linear algebra.

Putting all these things together and using results of Wall–Mishchenko, we obtain that the signature of the universal cover $\tilde{M}$, taken equivariantly with respect to the fundamental group, is in fact an element of the K-theory of the C^*-algebra of the group

$$\mathrm{Sign}_c(\tilde{M}) \in K(C^*(\Gamma)). \tag{16}$$

The problem is as follows. If we were in the Abelian case, then this C^*-algebra would be the continuous functions on the Pontrjagin dual of Γ, and the next step would be trivial; it would just be to take the Chern character of this signature. (Of course, it would be nontrivial to compute this Chern character; here is where the Atiyah–Singer index theorem for families would come in. Nevertheless, there would be no need to define a new theory of signatures; we could just take the Chern character, and compute it.)

If the group is non-Abelian, we do not have the space. We would like to say that this C^*-algebra of Γ is like continuous functions on some space, but we do not have the Pontrjagin dual because the algebra can drastically fail to be Abelian. It turns out that what is needed in order to replace the Chern character is, first, to think about the theory of characteristic classes and to be able to understand the theory in such a way that it will still hold in the non-Abelian case. This gives *cyclic cohomology*, with which I will now deal. This theory is motivated very strongly by the example, in the sense that there is a need for a replacement for the calculations of curvature, characteristic classes, and so on, in this non-Abelian situation, where we cannot use the usual setting.

Cyclic Cohomology

Let me try to present cyclic cohomology as simply as possible. It is just a generalization of the notion of trace. If we have a noncommutative algebra $\mathcal{A}$, then there is a simple equality on a functional —on a linear form of this algebra— which enables us to erase the noncommutativity, i.e., which enables us to do many things as if the algebra were commutative. This is the notion of a *trace*

$$\tau : \mathcal{A} \to \mathbf{C}. \tag{17}$$

The trace satisfies the following cocycle condition:

$$\tau(a^0 a^1) - \tau(a^1 a^0) = 0. \tag{18}$$

A cyclic cocycle, in general, is just a higher trace. By *higher trace* I mean that it is again a functional, but on several variables in the algebra, and satisfying the following two conditions

$$\tau(a^0 a^1, a^2, \ldots, a^{n+1}) - \tau(a^0, a^1 a^2, \ldots, a^{n+1}) + \cdots$$
$$\cdots + (-1)^n \tau(a^0, a^1, \ldots, a^n a^{n+1}) + (-1)^{n+1} \tau(a^{n+1} a^0, a^1, \ldots, a^n) = 0, \tag{19}$$

$$\tau(a^1, a^2, \ldots, a^n, a^0) = (-1)^n \tau(a^0, a^1, \ldots, a^n). \tag{20}$$

A simple example of a cyclic cocycle appears in the situation where the algebra is the algebra of functions on a manifold. Assume given a de Rham current (recall that a *de Rham current of dimension k* is a linear form on differential forms of degree k). When I say that it is *closed* I mean that when it is paired with a closed form it yields 0.

If we start with a closed current c, then we can indeed define a multilinear functional on the algebra by the following formula:

$$\tau_c(a^0, \ldots, a^k) = \langle c, a^0 da^1 \wedge \ldots \wedge da^k \rangle, \tag{21}$$

and it is not difficult to show that it satisfies conditions (19) and (20) above. Condition (19) is just the fact that the differential of a product is given by the Leibniz rule, and condition (20) tells us that the current is closed, so we can integrate by parts in the current and this enables us to cyclically permute the variables.

In order to extend the previous functional to matrices by multilinearity, one simply has to extend it on tensor products of functions by matrices. There is only one natural formula that can be applied:

$$\tau'_c(a^0 \otimes \mu^0, a^1 \otimes \mu^1, \ldots, a^k \otimes \mu^k) = \tau_c(a^0, \ldots, a^k) \operatorname{Tr}(\mu^0 \cdots \mu^k), \tag{22}$$

where $a^0, \ldots, a^k$ are functions, $\mu^0, \ldots, \mu^k$ are $q \times q$ matrices, and Tr denotes the ordinary trace. Observe that this new expression is not invariant under all permutations, because the trace of a product is only invariant under cyclic permutations. It is precisely this small fact which forces us to consider only cyclic permutations.

Why are traces important? The trace on an algebra is important because the trace automatically gives a *dimension* to any finite projective module. If we have a finite projective module over an algebra, this module can be viewed as an idempotent in matrices. The trace extends to matrices, and when we evaluate the trace on the corresponding idempotent, it does not depend upon any choice.

A higher trace (i.e., a cyclic cocycle) gives us an invariant, exactly like the Chern character, for finite projective modules. We shall see by very simple examples that this reduces to the Chern character in the example of a current given above.

It turns out that the evaluation of a cyclic cocycle τ of even dimension on a diagonal element $\tau(e, e, \ldots, e)$, for $e \in \operatorname{Proj}(M_q(\mathcal{A}))$, is homotopy invariant. In other words, if we move the idempotent by deformation among idempotents, then this quantity does not change. How does one prove this? The point is the following: If you move an idempotent among idempotents, then, of course, it is a nice spectral deformation, because the spectrum of an idempotent is only formed by 0's and 1's, so there has to be a nice spectral equation to satisfy. This equation is

$$\dot{e}_t = [x_t, e_t] \tag{23}$$

for some element x_t. This equation is easily obtained by differentiating the equation $e_t^2 = e_t$. Now, when we differentiate $\tau(e, e, \ldots, e)$, we get an $\dot{e}$ appearing only once at a time, and then, by a little algebraic manipulation using the cocycle identity, we can prove that we get 0. So this is invariant under deformations, and, moreover, it is not difficult to prove that it only depends upon the isomorphism class of the finite projective module defined by e. Moreover, it is additive, so that if we take the direct sum of two finite projective modules —even if we have a monomial which is not linear— what we get is a sum of the corresponding traces.

This means that the so-called *cyclic cohomology*, where elements are cyclic cocycles modulo an obvious relation, pairs with K-theory, so each cocycle class defines a map from the K_0 of the algebra, $K_0(\mathcal{A})$, to the scalars.

Let me show with this example of currents, first, how this computation reduces to the Chern–Weil computation by connections and curvature for vector bundles. I showed that if we have a closed de Rham current on M, then we have a cyclic cocycle. Of course, if we want to know the Chern character pairing, it is enough to know how the Chern character pairs with any closed de Rham current, because the closed de Rham currents generate the homology of the manifold. So what we have to do is to show the equality

$$\langle \tau_c, [E] \rangle = \tau_c(e, \ldots, e) = \langle \operatorname{ch}(E), c \rangle, \tag{24}$$

where $[E]$ is the finite projective module of the vector bundle E, and c also denotes the homology class of the current. How does one prove this? The finite projective module of a vector bundle is given by an idempotent. A more geometric way of formulating this is to say that the vector bundle is the pull-back of the canonical vector bundle on the Grassmannian by a map from the manifold to the Grassmannian, because when we take an idempotent in $n \times n$ matrices over the algebra, just by a matter of translation, this is exactly a map from the space to the set of idempotents of $n \times n$ matrices, which is the Grassmannian. On the Grassmannian we have a canonical connection, which comes from the orthogonal projection from one fibre (i.e., from one vector space) to the nearby vector space. Now we can pull back this canonical connection.

If we compute the curvature of this connection, we will find that it is given as a matrix of differential forms $edede$, where de is the differential of this map e. And so, when we pair the curvature to some power with the current, we immediately see that we get $\tau_c(e, e, \ldots, e)$.

What we have done is to translate algebraically the pairing in such a way that, firstly, it is now completely free of the commutativity hypothesis; and, secondly, that it relates in fact to cohomology which is well defined, because, if one looks at the definition of a cyclic cocycle, one easily understands that condition (19) is the condition of being closed. Observe that the sum appearing in this condition is nothing other than the Hochschild coboundary of the cochain we are dealing with. And condition (20) is a restriction to cyclic cochains, which turns out to be stable under coboundary, so what we get is a complex, and out of this complex we get, of course, a cohomology theory which is cyclic cohomology.

The Equivariant Index

Let us look at other examples. Take the group ring of a discrete group, $C[\Gamma]$, and suppose given a group cocycle $c(g_1, \ldots, g_n) \in C$, $g_i \in \Gamma$.

Remember that when we were considering the Pontrjagin dual, we wanted to compute the Chern character. The problem was, of course, only to be able to pair this Chern character with group cocycles. Now we still have the group, but it is not Abelian, so we cannot talk about the Pontrjagin dual. However, we have the group ring, and the claim is that the following extremely simple formula

$$\tau_c(g^0, \ldots, g^n) = \begin{cases} 0 & \text{if } g^0 g^1 \cdots g^n \neq e \\ c(g^1, \ldots, g^n) & \text{otherwise} \end{cases} \tag{25}$$

assigns to every group cocycle a cyclic cocycle on the group ring.

Now the main trouble is that we do not have the Atiyah–Singer index theorem. Remember that when we were doing the calculation in the case of the Pontrjagin dual of an Abelian group, we used the Atiyah–Singer index. So we need a replacement for it. This replacement is a theorem which will not only handle the signature operator, but in fact will handle an arbitrary elliptic Γ-invariant operator on the covering space $\tilde{M}$.

If we are given a differential operator D which is elliptic on the manifold M, we can always lift it (because it is local) to the covering space, into an operator $\tilde{D}$ which is Γ-invariant and still elliptic. It turns out that, while an operator downstairs has a *parametrix* (an inverse modulo smoothing operators) the operator on the covering also has a parametrix, but this one is not an inverse modulo smoothing operators: It is an inverse modulo $R\Gamma$, the group ring of Γ extended by the smoothing operators. (R denotes the ring of smoothing operators on the base, which does not depend on the manifold.)

In fact, the index for the operator is an element of the K-theory of the group ring of Γ extended by the smoothing operators, $\mathrm{Ind}_\Gamma \tilde{D} \in K_0(R\Gamma)$, which is called Γ-*equivariant index*. Now the following theorem holds, which is exactly a higher analog of the Atiyah–Singer index theorem, in the same way as the Novikov higher signature is the analog of the ordinary signature.

THEOREM (CONNES–MOSCOVICI). *If c is a group $2q$-cocycle on the group Γ, then*

$$\langle \tau_c, \mathrm{Ind}_\Gamma \tilde{D} \rangle = \frac{1}{(2\pi i)^q} \frac{q!}{(2q)!} \langle \mathrm{ch}\sigma_D \cdot \mathrm{Td}(M) \cdot [c], [M] \rangle. \tag{26}$$

Here Td stands for the *Todd genus* of the complexified tangent bundle. This formula contains two new terms with respect to the Atiyah–Singer formula; namely, the numerical constant $q!/(2q)!$, which takes care of the dimension of the cocycle, and the factor $[c]$, which is the class of the group cocycle c viewed on the manifold M.

Now we apply this to the signature operator on the covering and get the following:

COROLLARY. *If D is the signature operator, then*

$$\langle \tau_c, \mathrm{Ind}_\Gamma \tilde{D} \rangle = \text{Novikov higher signature} = \langle L[M] \cdot \varphi^*(c), [M] \rangle. \tag{27}$$

We could say that now the Novikov conjecture is solved in general; but there is still one technical problem. (Nevertheless, the conjecture is solved for a generic family of groups, namely *Gromov hyperbolic groups*. I will not give the technical definition of these groups, but I will mention the technical reason which still restricts the proof to these groups.)

When dealing with the Abelian case, one has smooth functions, and smooth functions have the following two rather important properties. The first property is that the K-theory of the algebra of smooth functions is the same as the K-theory of continuous functions

$$K(\mathcal{C}^\infty(M)) \xrightarrow{\cong} K(\mathcal{C}(M)). \tag{28}$$

The second is that the cyclic cocycles that we get automatically extend from the group ring $\mathbb{C}\Gamma$, which is like Laurent polynomials, to smooth functions. It turns out that, when we take non-Abelian groups, then this problem —which is quite trivial in the Abelian case— has analytical difficulties. However, this technicality can be solved for Gromov hyperbolic groups.

I will now explain in what sense these groups are generic. By a result of Gromov, if we look at (finitely presented) groups given by generators and relations, pick a finite number of generators, bound the length of the relations, and count among the obtained groups those which are hyperbolic, then the percentage of these tends to 100% as length tends to infinity.

We would like to have the Novikov conjecture true in general. It might be that the above technical problem is indeed essential, and the conjecture is only true for groups for which one has a sort of analytical control. It is very important to deal with such questions, because it is not only for the Novikov conjecture that they are relevant. What one is really dealing with in this situation is analysis on the dual of a discrete group. This is a quantum space (it is not Abelian) and this analysis is much more complicated and rich than in the Abelian case. Essentially, the Abelian case is a sort of finite-dimensional case, while in the non-Abelian case, because of the growth of the groups, we have phenomena which are infinite-dimensional in nature.

Riemannian Geometry

At the beginning I explained the original motivation of Heisenberg. Then I showed by means of some examples how the idea of noncommutative geometry can be used in specific examples. In the foregoing discussion, I have been dealing only with topology, K-theory, differential forms, and characteristic classes. Now I want to discuss the very essential part of geometry that deals with the measurement of lengths, i.e., Riemannian geometry. This is by no means finished. Although only a small part of it is complete, I think it will have applications in physics, so I want to return to physics in the last part of my talk.

In doing noncommutative geometry, one arrives after many examples at the following notion of what might replace the notion of Riemannian manifold. This notion, on one hand, will cover the finite-dimensonal case; but in Riemannian geometry it will do something more: It will mix the discrete and the continuum. (Riemannian geometry, as it is usually known, deals only with the continuum and does not handle the discrete.) It will also make it possible to handle nonintegral dimensions, like Hausdorff dimensions. For instance, if we have a circle which is winding in a set of higher Hausdorff dimension in the plane, it will enable this to be handled exactly as it would be in Riemannian geometry, but the functions will not be differentiable.

Let me come back to the fundamentals of Riemannian geometry. Riemannian geometry deals with a certain metric space where the distance is computed as the infimum of the arc-lengths, $d(p,q) = \text{Inf}\{\int_p^q ds\}$, where the length of an infinitesimal arc is given by the square root of a quadratic form

$$ds = \sqrt{\sum g_{\mu\nu}dx^\mu dx^\nu}\,. \tag{29}$$

This geometry is amazingly relevant for two reasons. One is that it has a wide variety of examples, and the second is that many tools are available. In particular, all the tools of differential and integral calculus are available and make computations possible.

At first, Riemannian geometry was meant to be a generalization of Euclidean and non-Euclidean geometries, so there was the temptation of restricting it to extremely special spaces, like the ones in which rigid motion is possible. General relativity has shown that this would be a mistake, because in general relativity one is obliged to consider all possible spaces of a certain kind, and one is obliged to vary among them.

Let me now turn from this to a more algebraic standpoint. We take the algebra $\mathcal{A}$ of functions on the manifold M —no regularity is assumed— and this algebra is supposed to act on a Hilbert space. This Hilbert space is the space of L^2-spinors $\mathcal{H} = L^2(M, S)$. Moreover, I take as given the Dirac operator D; that is, what we are given is an algebra of functions together with a representation. But if we were just handling representations, we would have nothing to work with. I want to add some finiteness condition. This finiteness condition is given by the Dirac operator D, which is finite in the sense that its inverse is compact, or in the sense that its eigenvalues go to infinity. And it is compatible with the algebra of functions, in the sense that if we permute the Dirac operator with functions, they do not commute, but what we get is bounded. (The Dirac operator is not bounded.)

I will next show how to recover the manifold M, the geodesic distance $d(p, q)$ on M, the Riemannian volume, the integration of functions, the gauge potential, and the Yang–Mills action, out of the purely operator-theoretic data $(\mathcal{A}, \mathcal{H}, D)$. And this will be done in such a way that we will not be limited to Riemannian manifolds, but after a while will be able to handle discrete spaces as well.

Let me go very briefly through the way the manifold M is recovered. We have the algebra, yet we do not quite have the regularity. We recover the regularity by asking that the commutator be bounded. Then by closing we get the algebra of continuous functions. By the well-known duality between the algebra of continuous functions and the points, we recover the points as a compact topological space

$$M = \text{Spectrum of the } C^*\text{-algebra } \overline{a};$$
$$a = \left\{ a \in \mathcal{A} \mid [D, a] \text{ is bounded } \right\}. \tag{30}$$

Let us look at the distance, which is much more interesting. The usual formula for the distance is the infimum over all arcs. I will replace this formula with a formula which will give the same answer (i.e., the geodesic distance), but which will be dual; instead of considering arcs embedded in the manifold, I will consider coordinates. I want to measure the distance between two points as follows:

$$d(p, q) = \text{Sup} \left\{ |a(p) - a(q)| \mid \| [D, a] \| \leq 1 \right\}. \tag{31}$$

Let us check that this is true. When we compute the commutator $[D, a]$, we find that this is Clifford multiplication by the gradient ∇a of the function a. To say that this operator has norm less than or equal to one is precisely to say that, at each point, this gradient has a length less than or equal to one. By a simple argument, this is precisely to say that the function is Lipschitz for the geodesic distance, with Lipschitz constant equal to one:

$$\frac{\text{Sup} |a(p) - a(q)|}{d(p, q)} \leq 1. \tag{32}$$

Thus we immediately see that one inequality is indeed given. To get the other inequality, we just take the function which is the geodesic distance to a given point p. This function is Lipschitz, so we can put it on the right-hand side and we are done.

What we get here is the same geodesic distance as usual. However, the measurement has been different, and, in fact, when we are doing measurements —not of long lengths, but of very small lengths— we are perfectly unable to use a path. It could be said, for instance, that a photon has a trajectory which is a path going from one point to another point. Yet this is not true: The photon in quantum mechanics is a plane wave having a definite momentum, so that there is no path of a photon, and, in fact, we are not measuring the distance by the formula of infimum of arc-length, but precisely by the formula (31).

Having this formula does not account for much, because we need to be able to integrate functions. There is an analysis of the residue —that is, what is called the *Dixmier trace* of operators on the Hilbert space— which enables us to write down the volume form in the Riemannian case purely operator-theoretically from the Dirac operator:

$$\int_M f\,dv = \mathrm{Tr}_w(fD^{-p}),\tag{33}$$

where p is the dimension, i.e., the order of growth of the eigenvalues of D: $\lambda_n \sim n^{1/p}$. This is related to the Tauberian theorem, in the sense that if we take the functional Tr_w —the Dixmier trace, which is not the ordinary trace— then it is related to the residue of the zeta-function of the operator at the point 1.

This is a trace which was discovered by Dixmier in 1966. Essentially, his paper was never read: It remained completely hidden in the literature for a very long time; but from the work of Manin–Wodzicki and Guillemin I noticed that the residue of pseudo-differential operators was the same trace, except that the Dixmier trace exists in general. It is not particular to the case of differential operators, or the set up of a manifold. So it could be used in general to perform integration in this general Riemannian-theoretic situation. And now there is this quite amazing fact that the Hausdorff measure (for instance, on the boundary of quasi-Fuchsian groups) is also given precisely by the Dixmier trace, although we are now in the non-integral-dimensional situation.

Thus one constructs first the integration of functions, the distance, and then proceeds to construct gauge theory. To construct gauge theory, one uses the Dirac operator, defines connections, vector bundles, curvature, and so on.

Let us go to the key point. This gauge theory has exactly the same features as the ordinary gauge theory. In particular, it is only in dimension 4 that one has a general theorem which relates the second Chern class with the Yang–Mills action. This follows from a completely general theorem using the Dixmier trace, which, in fact, justifies the Dixmier trace and provides an inequality showing that the gauge theory is not trivial when the second Chern class is not trivial.

THEOREM. *Let $(\mathcal{A}, \mathcal{H}, D)$ be a triple with $D^{-1} \in \mathcal{L}^{2n+1}$. Then:*

1. *The equality $\varphi(a^0, \dots, a^{2n}) = \mathrm{Tr}_w(\gamma a^0 [D, a^1] \cdots [D, a^{2n}] D^{-2n})$ defines a Hochschild cocycle on $\mathcal{A}$.*

2. *The class of φ is the same as the class of the Chern character of the K-homology class of $(\mathcal{A}, \mathcal{H}, D)$.*

Now I would like to show what happens because of the fact that the theory is not limited to the continuum. We may consider a space which is a product space of a continuum (the ordinary four-dimensional continuum) by a discrete space, and the simplest discrete space we can take is a two-point space. One translates algebraically the meaning of taking a product by

$$A = A_1 \otimes A_2,$$
$$H = H_1 \otimes H_2,$$
$$D = D_1 \otimes 1 + \gamma_1 \otimes D_2. \tag{34}$$

Let us do gauge theory for this two-point space. The two-point space is described by the algebra $A = C \oplus C$, since functions on the two-point space are just given by two complex numbers $(f(a), f(b))$. What is the Dirac operator there? By a general theory which is called K-*homology*, it can be shown that it reduces to the following form: The Hilbert space is of the form $H = C^N \oplus C^N$; the algebra will act by the matrices

$$\begin{pmatrix} f(a) & 0 \\ 0 & f(b) \end{pmatrix} \tag{35}$$

and the Dirac operator D will be off-diagonal and, of course, self-adjoint:

$$D = \begin{pmatrix} 0 & M^* \\ M & 0 \end{pmatrix} \tag{36}$$

for a certain $N \times N$ matrix M. So this is the structure that we want to consider on the space.

Now, the first thing we have to do is to compute the distance. If we take for the distance the formula given by the infimum over the arc-lengths, then, since we have a two-point space, we will get nothing, because there is no arc in the two-point space. But we have the other formula, and we can compute the distance between our two points. Using the formula (31) we get

$$d(a, b) = \text{Sup}\left\{ |f(a) - f(b)| \ \Big| \ \| [D, f] \| \leq 1 \right\} = \frac{1}{\lambda}, \tag{37}$$

where λ is the norm of the matrix M, that is, the square root of the largest eigenvalue of M^*M. Then we compute the gauge potential, the Yang–Mills action, and we find a term that is precisely the so-called *symmetry-breaking term*, which physicists were obliged to introduce in order to assign masses to elementary particles.

Then we go a little further and ask ourselves what are vector bundles over a two-point space. Of course this is very trivial: A vector bundle is given by two fibres, C^k and $C^{k'}$, and a nontrivial bundle is one in which $k \neq k'$. We pick the simplest nontrivial bundle, which has fibres of dimension 1 in one point and of dimension 2 in the other point.

Once it is seen in detail what is the Riemannian case in dimension 4 and the two-point discrete case, we can look at the product case. When we take the product of these two spaces and compute what is the gauge theory, we find exactly what physicists have been given, too, by elementary particle physics in the so-called *Glashow–Weinberg–Salam model*. One finds a Lagrangian which comprises many more terms than the usual Lagrangian. Ordinarily, from Maxwell theory and Dirac theory, we know that the theory of quantum electrodynamics is described by one Lagrangian

$$\mathcal{L} = \mathcal{L}_B + \mathcal{L}_f + \mathcal{L}_{fB}, \tag{38}$$

which has a pure gauge potential part $\mathcal{L}_B$, a fermionic part $\mathcal{L}_f$, and an interaction between fermions and bosons given by the diagram which tells us how a photon can give a positron and an electron, for instance. This Lagrangian is that of quantum electrodynamics.

In this century, it has been understood that quantum electrodynamics was not enough to describe the so-called *electroweak interaction*. In fact, it has been discovered that there is a nuclear beta decay, that there is radioactivity (which was discovered at the end of the last century), and, gradually, with a lot of experiments, people have been led to the following experimental Lagrangian:

$$\mathcal{L} = \mathcal{L}_B + \mathcal{L}_f + V(H) + \mathcal{L}(B, H) + \mathcal{L}(f, H). \tag{39}$$

where $V(H)$ is the *Higgs potential*, $\mathcal{L}(B, H)$ is the *minimal coupling*, and $\mathcal{L}(f, H)$ is the *Yukawa coupling*.

By doing a small calculation in noncommutative Riemannian geometry, I have shown that if one alters the space a little bit by crossing it with a discrete set of two points, then space-time becomes like a product of ordinary space-time by two points, and these two points are extremely close: If one computes their distance, one finds something like 10^{-16} cm.

The idea consists of not just introducing new dimensions, but to pick a discrete fibre. Now, when we compute the Lagrangian as explained above, for this new Riemannian space we find exactly the standard model with all its five terms (39).

At the moment, in order to incorporate quarks, one has to do a little more. There are two copies of the space, i.e., there are two sides: one is *left-handed* and the other is oriented the other way. In order to incorporate quarks, instead of considering only scalar-valued functions on the left-handed copy, we have to consider quaternionic-valued functions. The algebra of quaternions is slightly noncommutative (by "slightly noncommutative" I mean that they satisfy polynomial identities; they are not something which is of high dimension with respect to matrices).

The general idea is that in order to understand space-time, it may be important not to be limited to ordinary Riemannian connected manifolds and to allow a more general notion of space-time —a more general notion of Riemannian geometry— based on operator-theoretic data and which makes it possible to talk about "effective" space-time. I am by no means saying that this is the final answer on space-time. What I am saying is that if we take this space-time and compute the analog of quantum electrodynamics on it, then we get precisely the complicated Lagrangian above (39). So this gives us a better geometric understanding of the finest existing effective model of elementary particle physics.

Alain Connes
Institut des Hautes Études Scientifiques
35 Route de Chartres
F–91440 Bures-sur-Yvette
France

Transcribed from the videotape of the talk by Pere Ara and Carles Broto; revised by the author.

Symposium on the Current State
and Prospects of Mathematics

Barcelona, June 1991

Theory of Computation

by

Stephen Smale

Fields Medal 1966

*for his work in differential topology, where
he proved the generalized Poincaré conjecture
in dimension n ≥ 5: Every closed n-dimensional manifold homotopy equivalent to the
n-dimensional sphere is homeomorphic to it.
He introduced the method of handle-bodies to
solve this and related problems.*

Abstract: It could be said that the modern theory of computation began with Alan Turing in the
1930's. After a period of steady development, work in complexity, specially that of Steve Cook and
Richard Karp around 1970, gave a deeper tie of the Turing framework to the practice of the machine.
I will discuss an expansion of the above to a theory of computation and complexity over the real numbers
(joint work with L. Blum and M. Shub).

Theory of Computation

The reason for a theory of computation, for me in particular, comes from an attempt to understand algorithms in a more systematic way. The notion of algorithm is very old in mathematics; it goes back a couple of thousand years. Mathematicians have talked about algorithms for a long time, but it was not until Gödel that they tried to formalize the notion of algorithm. In Gödel's incompleteness theorem one saw for the first time the limitations of computations or the need to study more clearly what could be done.

To do so, one has to establish more explicitly what an algorithm is, and I think that this became clearer in the way that Turing interpreted Gödel. So let us stop for a moment and look more closely at Turing and his achievements. I think it is fair to say that he laid down the first theory of computation. Perhaps I will be more specific later about what Turing's notion of computation was.

We can take as set of inputs the integers $\mathbf{Z}$, so in the Turing abstraction the input is some integer; perhaps not every integer is allowed but only those that were eventually called the *halting set* Ω_M of the machine M. This is the domain of computation of the machine M. Given an integer in Ω_M, we feed it to the machine M and obtain as output another integer.

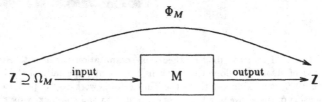

There is some kind of mechanism here, described by Turing, which I will later on formalize in my own way. Turing gave different versions of the input set; for instance, finite sequences of zeroes and ones.

Thus Gödel's incompleteness theorem can be stated in the following way.

THEOREM. *There is some set $S \subset \mathbf{Z}$ which is definable in terms of a finite number of polynomial conditions and is not decidable.*

This is Gödel's incompleteness theorem as formulated by Turing. *Not decidable* means that there is no computable function over $\mathbf{Z}$ which is 1 on S and 0 out of S. In other words, S is *decidable* if its characteristic function is computable by some machine. Thus, Gödel's incompleteness theorem asserts that there exists a set S which is very definable mathematically, yet is not decidable.

This is in some sense the beginning of the theory of computation, which shows the limits of decidability. Eventually, from this evolved a theory for present-day computers. It is from this formulation that it evolved into one of the foundations of computer science. Even a very refined theory of computer science is developed from this: This is complexity theory, which I could say today lies in the center of theoretical computer science, specially after the work of Cook [3] and Karp [7]. They made use of the notion of speed of computation; now the question is not whether a set is decidable, but whether it is decidable in a time that can be affordable by present-day machines, or whether it is a "crackable" problem.

The fundamental question is

$$P \neq NP?$$

This is a very famous conjecture and it is the most important new problem in mathematics in the last half of this century; it is only 20 years old. To me it is the most beautiful new problem in mathematics. Very hard to solve, a very fine notion coming from this theory of Cook and Karp.

So we have a very active subject in this area but there is something that is missing. I have talked about the need for a notion of definable algorithm, yet the algorithms mathematicians have used for a couple of thousand years at least do not fit into this framework. The algorithms we are talking about have to do with real numbers, and specially since the time of Newton they have had to do with differential equations, nonlinear systems, etc. We see the notion of the real numbers **R** is central.

Newton's method to me is a paradigm of a great classical algorithm like the procedure of the Greeks for finding square roots, and it does not fit naturally into this framework, because the framework is quite discrete and to fit Newton's method into it requires destroying geometric concepts. One can do this in a very cumbersome way —I find this a very destructive way— to deal with the algorithms of continuous mathematics with Turing machines.

Indeed, some work has now been done to adapt the Turing machine framework to deal with real numbers. Let me mention two such attempts. One of them is recursive analysis, which initially was worked out by Ker-I-Ko and Harvey Friedman [5] and the main name connected with it is Marian Pour-El. She worked with Richards [8] in developing a kind of real number analysis based on Turing machines. There has been very extensive work on this which deals with partial differential equations, and the way to deal with real numbers in this context is to consider a real number s defined by its decimal expansion

$$s = 1.2378\ldots.$$

A real number is computable in this sense if there exists a Turing machine which says that the first digit is 1, the second 2, the third 3, and so on, with the decimal point in the appropriate place. So a computable real number is given by a Turing machine. These work with computable real numbers and eventually provide a very successful theory. Similarly there is the notion of *interval arithmetic* from R. E. Moore (see [1]). In some ways it is close to the work of Pour-El but in a quite different direction. Thus, the foundations are probably being laid for a theory of computation over the real numbers.

Now, continuing from a very different point of view, I will devise a notion of computation taking the real numbers as something axiomatically given. So, a real number to me is something not given by a decimal expansion; a real number is just like a real number that we use in geometry or analysis, something which is given by its properties and not by its decimal expansion. Eventually, I will talk about a notion of computability over the real numbers which takes this point of view. There one thinks of inputing a real number not as its decimal expansion but as an abstract entity in its own right.

Some mathematicians and computer scientists have trouble with the idea that a machine takes as input an arbitrary real number. I wrote a paper [13] on precisely this point, saying that here one idealizes, as in physics Newton idealized the atomistic universe —making it a continuum— in order to use differential equations. One can idealize the machine itself by conceiving it as allowing an arbitrary real number as input, but I am not going to argue about this point today.

In a preliminary phase, I was concerned with the problem of root finding for polynomials for many years. In that process I faced the kind of objects known as *tame machines*. It is not a theory of computation, but just a preliminary.

We take as input now the coefficients $\{a_0, a_1, \ldots, a_d\}$ of a complex polynomial f of degree d, and we think of it over the real numbers, i.e., each a_i is given by its real and imaginary parts. Thus, we think of this as the input and we describe the computation in the language of flowcharts. Then comes a box describing the computations. We replace a by $g(a)$, where g is a rational function. This vector of numbers $[a_0, a_1, \ldots, a_d]$ can be considered as a state and in this step this state is transformed by a rational function. Then we can put down and answer the question of whether some coordinate of the state is less than or equal to 0. Depending on the outcome of this comparison, we continue along the corresponding branch.

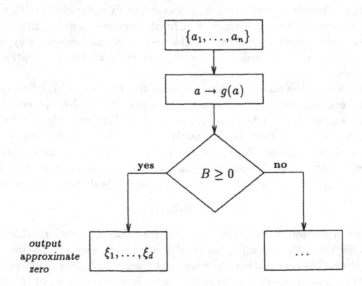

So we go down the tree in this way, and eventually we may output the approximate zeroes $\{\xi_1, \ldots, \xi_d\}$ of f up to some ε.

In fact this is a good way to express an algorithm for solving this equation. We next ask: How about these nodes which branch? To what extent are they necessary?

The topological complexity of a problem in general is the minimum number of these branch nodes for any machine which solves the problem.

I am not being completely precise about what resolving a problem means. One can imagine this example as a prototype of the general situation of a machine that solves problems. In particular, for this problem of finding the zeroes of polynomials we have the notion of its *topological complexity*. And the theorem [12] is as follows:

THEOREM. *The topological complexity of the root finding problem is greater than or equal to $\log d$.*

So the topological complexity increases with the degree, and the proof of this theorem actually is not so easy; it uses the cohomology of the braid group worked by Fuchs in Russia [6]. Subsequently, Vasiliev [14] extended this bound to $\simeq d$. So the answer eventually emerged that the topological complexity grows linearly with d and this is a sharp bound.

One notes that the Turing machine framework could never deal with this way of looking at all possible algorithms, even in this limited class. There is no useful way of thinking about it in terms of Turing machines, whereas using this kind of tree we were able to give necessary conditions on all algorithms, what we call *lower bound theorems*. There is some early work dealing with this kind of tree, but this is the first time we have obtained topological complexity results using algebraic topology.

Then, shortly after this, we did a joint work with Lenore Blum and Mike Shub [2] and developed this into a complete theory of computation over the reals by allowing loops. We certainly increased the computational power of tame machines by allowing loops to give a notion of computation in general. This situation is reflected in the next picture.

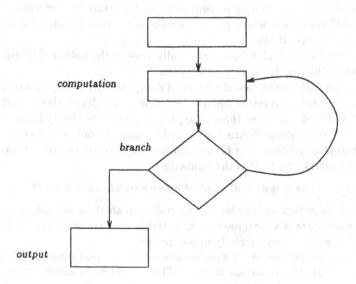

So here we have also

- an input space $\mathbf{R}^l$,

- an output space $\mathbf{R}^k$,

- and also a state space $S = \mathbf{R}^j$ for things happening inside the machine.

If l, k, j are finite, this essentially defines a machine. We can take here an oriented graph where the nodes are computation nodes given by rational functions, branch nodes given by inequalities, and input and output defined inside accordingly. At each computation node there is a single output, a single branch going out of the node. A decision node has two. Remember that the number of input branches is arbitrary except for the input node where nothing comes in and there is a branch going out, and an output node, where nothing comes out.

This gives a theory of computation for finitely dimensional input and output spaces motivated directly by the flowcharts used in scientific computation. Yet the full theory will have to allow $l, k, j = \infty$ and we will have to have a little more technical process to access far out coordinates, but this is the idea.

This model gives an algebraic flavour to the process of computation. We defined this not only over the real numbers but over any ordered ring, eventually any ring. In particular, if we take the ring to be $\mathbf{Z}$, the input space to be a subset of $\mathbf{Z}$, the output space again $\mathbf{Z}$, and the state space $\mathbf{Z}^\infty$, we obtain Turing theory, and so this extends the Turing theory of computation.

We can now say that a Turing computable function is one which is given on some Ω of the machine, a domain of inputs, by following the flow of the machine and doing what is said at each node.

$$\{admissible\ inputs\} = \Omega_M \xrightarrow{\Phi_M} \mathbf{R}^k$$

And this essentially is a complete picture of what we mean by *computable function*. A function Φ_M defined by a machine going from the admissible inputs or the halting set of the machine to the output set.

And it is precisely equivalent —or practically so— to the notion of *Turing computable* in the case when the ring is the ring of integers.

We have developed for this model notions of computability itself; we have, for instance, shown the existence of universal machines. We have a complexity theory and the problem "$P \neq NP$?" is also defined over these rings; for example, the theory for $\mathbf{R}$ or $\mathbf{C}$ possesses universal or NP-complete problems just as in the case of Cook and Karp.

An NP-complete problem over $\mathbf{C}$ (a machine over $\mathbf{C}$ is like one over $\mathbf{R}$ except that the branch nodes just ask "$\neq 0$?") is the following:

Does a system of quadratic polynomials have a zero?

The idea of the reduction is to have more polynomials than variables. So it is an open question whether there is a machine that can decide in polynomial time if there is such a zero. All this is written very carefully in our paper.

In Barcelona, Felipe Cucker [4] gave an analog for the real numbers of the arithmetical hierarchy of classical recursion theory. There have been developments in different directions in the theory of computation of these machines from the point of view both of complexity theory and of computability.

There has also been a lot of controversy and criticism. Let me deal with one main point, making some comments on two sharp critiques by Pour-El and Moore. Our theory of computation is very different from their two theories. In a way this is more or less the basis of their criticism. It has to do with the branching

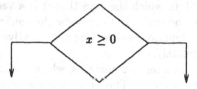

We branch according to whether one of the coordinates of the state space is greater than or equal to 0. This is in some respects one of the most controversial elements in the kind of machines we have, because the question is that an actual machine cannot do this. Given a number, for example

$$0.0000\ldots00\ldots,$$

it may or may not have a one after that eventually. If it never has a one, and we input it to an actual machine, we can never decide this question. If it does have a one, we wait long enough and we can decide it.

So we have a problem here when branching at ≥ 0, or equivalently at $= 0$, and this is the focus of attack of both Pour-El and Moore. Let me give an example here.

Both of their theories of computation lead to a notion of computable function which is continuous. Every computable function here is continuous. Even in a strong sense: They have to be constructively continuous.

Now the clue to this lies in the philosophy of thinking about the real numbers as abstractly given, and choosing the idealization of the right machines. For example, in scientific computation this is the kind of computation carried out traditionally by algorithms like Newton's method. One does test if something is ≥ 0, then do this, if not do something else.

Moreover, the need for these branchings is given by our earlier results on topological complexity. Topological complexity states that if one wants to find zeroes of polynomials then one has to branch, and the number of branchings in the machine is given approximately by the degree. Even to approximately solve the fundamental theorem of algebra one needs to branch. So I would imply that these two theories of computation do not lead even to an approximate solution of the fundamental theorem of algebra.

Here I would refer to a letter I received from Moore a year ago. I do not intend to dwell here on my opinion that numerical analysis and scientifical computing have weak foundations. Moore is the main developer of interval arithmetic and he wrote that "There are foundations for scientifical computation. More than 2000 papers and dozens of books. I invite you to read all of Aberth's book" [1]; it is a book that Moore even sent to me. He said "It will open your eyes to a whole new world." So I opened the book —actually a few weeks ago— and read on page 34 of the book (called *Precise Numerical Analysis*) "The problem of deciding whether two computable real numbers are equal is therefore a computational problem one should avoid." But problem 3.1 on this book reads: "Given two numbers a, b decide whether $a = b$." Later, on page 62, Aberth says: "Solve the problem 6.1: Find k decimals for the real and imaginary parts of the zeroes of a polynomial of positive degree." But the answer to this solvable problem —the fundamental theorem of algebra— needs to pass through d versions of this single problem that "one should avoid."

Marian Pour-El very kindly sent me a review that she has given of our paper in the Journal of Symbolic Logic [9], in which she says that it is a very good, highly developed theory of computability over the reals. In the review she confirms that in her theory she can only produce continuous functions and so she cannot solve the fundamental theorem of algebra, not even approximately.

What I will now do is to pass on to something which relates to this problem of NP-completeness if only a little indirectly. This is work done jointly with Mike Shub in the last few months. It is an example of an algorithm which fits into our framework. But it is a simple algorithm, so the fact that it is an algorithm in our strict sense is secondary. It is the problem of the complexity analysis of Bézout's theorem. Let me say a little bit about what this is. The situation we look at is as follows: We have a polynomial system

$$f : \mathbf{C}^n \to \mathbf{C}^n$$

of n polynomials in n variables of degrees $d_1, \ldots, d_n$ respectively. One wants to find an algorithm and analize its speed for solving the equation

$$f(z) = 0 \,;$$

not to produce a solution but to analize how much time it takes. The idea is to make complexity analysis on this.

The work done so far on polynomial equation solving can be summarized by dividing it into two parts: one is Newton's method as the basic algorithm —it is essentially the method used by the Greeks for finding square roots— and the other method is elimination theory, a very algebraic method; it works over arbitrary fields. In the first one we are using some kind of norm or metric, so it is metric-oriented. It is the method of choice of numerical analysts. The second is probably the method that would be chosen by a computer scientist. My own inclination is to the first side. Numerical analysts have a better focus on the problem. They do not have a complexity theory or any kind of foundation, but they have a better instinct about how to solve this problem.

In any case, what we use is some global version of Newton's method to solve Bézout's theorem. That is, we use Newton's method to follow a path in the space of polynomials.

All these are homogeneized. It is more elegant to think entirely in terms of homogeneous coordinates and projections

$$\mathcal{P}_{(d)} = \{f : \mathbf{C}^n \to \mathbf{C}^n\}$$

and

$$\mathcal{H}_{(d)} = \{f : \mathbf{C}^{n+1} \to \mathbf{C}^n \text{ homogeneous of degree } d\} \,,$$

with

$$\mathcal{P}_{(d)} \equiv \mathcal{H}_{(d)} \,.$$

Thus we are going to work in projective space, and the main thing we will analize is a projective version of Newton's method which is due to M. Shub [11]. The previous work has been done mostly in one variable on this problem, using Newton's method to solve this; I spent many years doing that. Jim Renegar [10] has some extension to n variables.

What we want to do here is give a very conceptual process. All the ideas are lying there; we can try to understand the best, most elegant ways of looking at algorithms to solve this. In this way perhaps eventually we will be able to see more clearly the problem "$P \neq NP$?" over the real numbers or the complexes. I hope it will eventually shed some light on the practical problem "$P \neq NP$?"

Given the space $\mathcal{H}_{(d)}$ we consider a function f_0 for which we know the zeroes. The zeroes of f_0 could be given by a set of intersections in a grid so we get a set of equally spaced zeroes

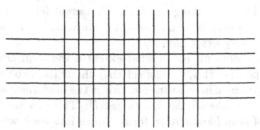

This could be the initial element in $\mathcal{H}_{(d)}$ for which we know the answer, and we simply homotope that back

$$f_t = tf + (1-t)f_0,$$

where t goes from 0 to 1. Let us denote by $\mathcal{F}$ the curve f_t in the space $\mathcal{H}_{(d)}$.

We try to trace the zeroes, which we know for f_0, to give the answer for f_1. This seems to be a very good method that has been used for the last decade or two. It embodies some kind of global Newton's method. The idea is to consider some sequence t_i and apply Newton's method to the function $f_{t_{i+1}}$, starting from some approximation X_i of the solutions for f_{t_i}

$$N_{f_{t_{i+1}}}(X_i) = X_{i+1}$$

where, here, $N_f(X)$ stands for applying Newton's method to solve f starting with the initial guess X. In the projective space $\mathbf{P}_n(\mathbf{R})$ we can see the paths given by the solutions X_t of f_t and the algorithm provides a sequence of points X_i following this path very closely.

So, what kind of results can we expect here? What kind of things may we prove? Here is the main theorem. The question is how many iterative steps are necessary, how many t_i, in such a way that we can follow this path very closely, and our result is

THEOREM. *The number of steps is bounded above by*

$$\frac{\hat{\alpha} D^{3/2}}{\rho^2}$$

where $\hat{\alpha}$ is a universal constant (given by a set of equations which can be solved itself by Newton's method) which is approximately $1/16$, D is $\max\{d_1, \ldots, d_n\}$ and ρ is the distance from the arc joining f_0 and f_1 to the discriminant variety.

It should be recalled that the discriminant variety is the subset of $\mathcal{H}_{(d)}$ of all singular polynomial systems. It is the variety of polynomial systems which are degenerate at some zero. And this is an algebraic variety that we shall call Σ. The theorem says that what is crucial are not the coefficients of f. They do not even enter. In fact, not even the dimension comes directly here; this is even dimension-free. But what is crucial here is the distance ρ between $\mathcal{F}$ and the discriminant variety Σ. This is the crucial factor —the only factor— in estimating the complexity for finding the zeroes of a polynomial system. Now we have to make a little caveat here because we are not finding the zeroes of every polynomial system. There may be a continuum of zeroes and then we cannot do this. So we have to put some kind of condition, let us say to solve $f + \varepsilon$ where ε is a small polynomial. This is the thing we solve. We cannot find the solution of arbitrary polynomial systems; there may be a continuum of solutions, but for some deformation we can find the zeroes in a very exact sense.

Now, the great problem to me is: To what extent is the term $D^{3/2}$ necessary?

While we have no proof of this, we suspect that the D itself could be eliminated from the formula. For a polynomial in one variable, this is the fundamental theorem of algebra, and we show that we can take off the 3/2 to get D. This is what we have done in the last months; the proof is in handwritten form. Since last week we believe that we can eliminate the D in the one variable case, but this uses the theory of Schlicht functions, which is only available for one variable. There is no theory of Bieberbach conjecture for more than one variable. If it is true, if it is D-free, if we can do this, then one can find for example one zero of a polynomial in one variable in a universal number of steps, say one hundred.

References

[1] O. Aberth, *Precise Numerical Analysis*, Brown Publishers, Dubuque, Iowa, 1988.

[2] L. Blum, M. Shub and S. Smale, On a theory of computation and complexity over the real numbers: NP-completeness, recursive functions and universal machines, *Bull. Amer. Math. Soc. (N.S.)* 21 (1989), no. 1, 1–46.

[3] S. A. Cook, The complexity of theorem-proving procedures, Proceedings 3rd ACM STOC (1983), 80–86.

[4] F. Cucker, The arithmetical hierarchy over the reals, to appear in *J. Logic Comput.*

[5] H. Friedman and K. Ko, Computational complexity of real functions, *Theoret. Comput. Sci.* 20 (1986), 323–352.

[6] D. Fuchs, Cohomologies of the braid group mod 2, *Functional Anal. Appl.* 4 (1970), 143–151.

[7] R. Karp, Reducibility among combinatorial problems, in *Complexity of Computer Computations*, R. Miller and J. Thatcher (eds.), Plenum Press, New York, 1972, 85–104.

[8] M. B. Pour-El and I. Richards, Computability and noncomputability in classical analysis, *Trans. Amer. Math. Soc.* 275 (1983), 539–560.

[9] M. Pour-El, Review of [2], to appear in *J. Symbolic Logic.*

[10] J. Renegar, On the efficiency of Newton's method in approximating all the zeroes of a system of complex polynomials, *Math. Oper. Res.* **12** (1987), 121–148.

[11] M. Shub, Some remarks on Bézout's theorem and complexity theory, to appear in *Proceedings of the Smalefest*, M. Hirsch, J. Marsden and M. Shub (eds.).

[12] S. Smale, On the topology of algorithms I, *J. Complexity* **3** (1987), 81–89.

[13] S. Smale, Some remarks on the foundations of numerical analysis, *SIAM Rev.* **32** (1990), no. 2, 211–220.

[14] V. Vasiliev, Cohomology of the braid group and the complexity of algorithms, to appear in *Proceedings of the Smalefest*, M. Hirsch, J. Marsden and M. Shub (eds.).

Stephen Smale
Mathematics Department
University of California
Berkeley, California 94720
USA

Transcribed from the videotape of the talk by Felipe Cucker, Francesc Rosselló and Álvaro Vinacua; revised by the author.

Symposium on the Current State
and Prospects of Mathematics

Barcelona, June 1991

Knots in Mathematics and Physics

by

Vaughan F. R. Jones

Fields Medal 1990

*for the discovery of a 1-parameter represen-
tation of the classical braid groups into a
family of C^*-algebras, which led to a polynomial invariant of links. His work has greatly
stimulated the recent development of research
in C^*-algebras, statistical mechanics, representations of quantum groups, knot theory
and invariants of 3-dimensional manifolds.*

Abstract: Mathematical knot theory began in the nineteenth century as a result of questions about
fluid flow and electromagnetism. In the twentieth century, it became a part of topology, the study of
smooth deformations.

In recent years, knot theory has gone back to its physical origins in an interplay of mathematical
analysis, topology, statistical mechanics and quantum field theory. It is far from clear what will emerge
from these developments.

Knots in Mathematics and Physics

The mathematical study of knots began in physics in the nineteenth century. The names associated with the first work on knots are those of Gauss and Listing. Gauss gave an integral formula for the linking number of two knots in 3-space. Later, Maxwell interpreted Gauss's formula when he was developing his theory of electromagnetism. The linking number of two knots was seen by him as the energy required to move an electric charge in a knotted charged wire complement along the other knot. Simultaneously to Maxwell, the other pioneers were Tait and Kelvin. Kelvin constructed his theory of *vortex atoms*, according to which the matter is constituted by small knots formed by something like vortex lines of smoke. The aim of Tait was to construct a table of knots. With only experimental methods, it took him ten years to get a table of the knots up to ten crossings, remarkably accurate in spite of some mistakes.

In the twentieth century, knots have been studied with the help of topology. Invariants of knots like the Alexander polynomial were obtained by means of the fundamental group of the knot complement in S^3. These simple invariants would have greatly simplified the hard work of Tait.

The Alexander polynomial $\Delta(t)$ was discovered in 1928 and can distinguish, for instance, the first three knots of Tait's table: the unknot, the trefoil and the figure eight knot, depicted below.

knot	$\Delta(t)$
	1
	$t^{-1} - 1 + t$
	$-t^{-1} + 3 - t$

The Alexander polynomial can be explained in terms of elementary topology of the knot complement in S^3. The variable t can be interpreted as the meridian of the knot. In 1934, Seifert answered two natural questions about the polynomial:

1. *Is there a knot which is not the unknot but has trivial $\Delta(t)$?* The answer is yes, and an example is the following knot:

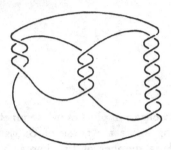

2. *Given a polynomial $P(t)$, is it the Alexander polynomial of some knot?* The answer is yes, provided that $P(t) = P(t^{-1})$ and $P(1) = 1$.

Via surgery, knots and links come into the realm of 3-manifolds. To do *surgery* on a link means to remove a tubular neighborhood of the link and to reglue it by means of a homeomorphism of the boundary. It was proved by Lickovish and Wallace that any closed orientable 3-manifold can be obtained by surgery on a link in the 3-sphere. A calculus for surgery was then developed by Kirby and, more recently, by Rourke and Fenn.

One can say that knot theory was more or less a quiet field when, in 1984, a change came with the discovery of a new polynomial $V(t)$. It had the same aspect as the Alexander polynomial, but it turned out to be, for instance, $t + t^3 - t^4$ for the right hand trefoil knot and $t^{-1} + t^{-3} - t^{-4}$ for the left hand trefoil! It is able to distinguish in many cases a knot from its mirror image. It is definitely different from the Alexander polynomial. It is not, at this moment, understandable in terms of algebraic topology.

At this stage, no nontrivial knot has been found with trivial polynomial $V(t)$. The two questions answered by Seifert for the Alexander polynomial have not yet been answered for $V(t)$.

There are at least four ways in which knot theory is related to physics.

1. Statistical mechanics.

2. Conformal field theory.

3. Topological 3-dimensional quantum field theory.

4. Algebraic quantum field theory.

There are relations between these fields. All have to do with braids, so let us introduce braids. A *braid* is a set of strings suspended from some vertices, like for the following

3-string braid:

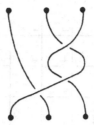

Braids of a fixed number n of strings naturally form a group, denoted traditionally by B_n. The operation is concatenation (it is not commutative). The *closure* of a braid is obtained by joining corresponding vertices, as in the figure:

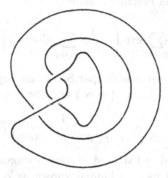

Alexander proved that all knots and links are obtained as braid closures. Braids that close to the same link are related by what is called *Markov equivalence*. Natural generators for B_n are the elementary braids σ_i

Artin proved that the relations $\sigma_i\sigma_{i+1}\sigma_i = \sigma_{i+1}\sigma_i\sigma_{i+1}$, $i = 1,\ldots,n-1$, and $\sigma_i\sigma_j = \sigma_j\sigma_i$ if $|i - j| \geq 2$, give a presentation for B_n.

Statistical Mechanics

In order to study critical phenomena in statistical mechanics, it is important to have available a certain number of mathematical models for which some thermodynamic quantities can be calculated exactly. Very few such models exist, and those only for 2-dimensional classical models in equilibrium on a lattice. The simplest of these is the square lattice Ising model solved by Onsager in 1944. One places "spins" taking the values ± 1 on the vertices of the lattice, and a state of the Ising model is given by the assignment of a spin value to each spin as indicated:

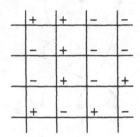

The spins are supposed to interact with their nearest neighbours, so there is an energy matrix $E(\sigma, \sigma')$, where σ, σ' are ± 1. The total energy of a state is the sum of $E(\sigma, \sigma')$ over the edges, and the partition function is then

$$Z = \sum_{\text{states}} \exp\left(-\frac{1}{kT} \sum_{\text{edges}} E(\sigma, \sigma')\right) .$$

This is defined for some finite region with specified boundary conditions. Then one lets the finite region fill up the whole lattice ($\rightarrow \infty$) and the free energy per unit site is

$$F = \lim_{\text{region} \rightarrow \infty} \frac{1}{\#(\text{spins in region})} Z .$$

This is what Onsager calculated and showed to have a singularity at the critical point.

Baxter solved more models by a technique known as *commuting transfer matrices*. With this solution, he showed that a certain "universal behaviour" as the system approaches criticality, which was true in the Ising model, was not true in general.

The idea of the transfer matrix technique is to build the lattice up row by row by inventing a matrix $T(\lambda)$ —where λ is any parameter involved, e.g. the temperature— such that $T(\lambda)^n$ corresponds to the lattice with n rows, in the sense that its entries are the partition function for a system with boundary conditions indexing the entry of the matrix. It is possible to write $T(\lambda)$ as a further product $(R_1(\lambda)R_3(\lambda)R_5(\lambda)\cdots)(R_2(\lambda)R_4(\lambda)R_6(\lambda)\cdots)$, where R_{odd} corresponds to the addition of a horizontal edge and R_{even} corresponds to addition of a vertical one. The relations

$$\begin{aligned} R_i(\lambda)R_{i+1}(\mu)R_i(\lambda - \mu) &= R_{i+1}(\lambda - \mu)R_i(\mu)R_{i+1}(\lambda) \qquad (*) \\ R_i(\lambda)R_j(\mu) &= R_j(\mu)R_i(\lambda), \qquad |i - j| \geq 2, \end{aligned}$$

then imply that the transfer matrices commute for different values of λ. This helps Baxter diagonalize $T(\lambda)$ and thus compute $T(\lambda)^n$ in the limit.

Note the similiarity between the equations $(*)$ and the braid relations. If λ were not present, these would be Artin's braid group presentation! It turns out that one may take a limit in λ such that the braid equations are nontrivially satisfied. Taking the trace of the resulting braid group representations gives knot invariants; in particular, the polynomial $V_L(t)$ comes for the "G-vertex model" defined by Baxter.

Quantum groups were invented to obtain solutions of the equations (∗). There results a picture in which there is a knot polynomial for every finite-dimensional representation of every simple Lie algebra. One may even treat the components differently by choosing a different representation per component. The rôle of the knot polynomials in the models remains quite mysterious.

Conformal Field Theory

At the critical point, statistical mechanical systems are supposed to have a quantum field theory in the continuum limit. Polyakov proposed that a statistical mechanical system at its critical point should exhibit "local scale invariance," i.e., if $\psi(x)$ is a field in the continuum limit, then $\psi(x)$ should transform in a simple way under transformations which scale the metric at each point x. An analysis of the quantum version of this scaling group in two dimensions yields the existence, on the Hilbert space of the field theory, of two commuting representations $L_n, \tilde{L}_n$ of the *Virasoro algebra* with generators c, L_n and relations

$$[c, L_n] = 0, \qquad [L_n, L_m] = (n - m)L_{n+m} + \frac{\delta_{n,-m}^{n^3-n}c}{n}.$$

The L_n's are infinitesimal generators of conformal transformations represented on the Hilbert space (and complexified).

These two representations allow one to decompose the Hilbert space into "sectors" where the L_n's and $\tilde{L}_n$'s act irreducibly according to some representation (on the "holomorphic" and "antiholomorphic" parts, respectively). The whole Hilbert space being irreducible, one expects fields $\psi(z)$, $\tilde{\psi}(z)$ intertwining the various representations of the Virasoro algebra, according to Seiberg–Moore. If one takes the vacuum expectation values, one obtains a function (possibly multivalued) on the space $\mathbf{C}^n \setminus \Delta$, where $\Delta = \{(z_1, \ldots, z_n)$ such that $z_i = z_j$ for some $i \neq j\}$. It is expected that the functions are indeed multivalued, being sections of some bundle on $\mathbf{C}^n \setminus \Delta$ with a flat connection whose monodromy gives representations of the braid group.

Following the work of Knizhnik and Zamolodchikov, Tsuchiya and Kanie calculated this monodromy for the so-called *WZNW model*, which is not obtained from statistical mechanics, but rather begins as a field theory, the fields in the classical case being functions from S^2 to some compact Lie group. The algebra of symmetries is larger than the Virasoro algebra and contains two copies of the affine Lie algebra of the compact Lie group (the central change being determined by a parameter in the Lagrangian of the classical theory). In this case, the connection on the bundle referred to above is the *Knizhnik–Zamolodchikov equation*, and Tsuchiya and Kanie calculated the monodromy in some cases for $SU(2)$ and obtained precisely the braid group representations required to calculate $V\left(e^{\frac{2\pi i}{k+2}}\right)$, k being the central change.

Topological Quantum Field Theory

Witten proposed the following formula for V:

$$V_L\left(e^{\frac{2\pi i}{k+2}}\right) = \int_A (\mathcal{D}A) \exp\left(ik \int_{S^3} \text{tr}(A \wedge dA + \frac{2}{3}A \wedge A \wedge A)\right) \cdot \prod_{C_j} P \exp\left(\int_{C_j} A\right),$$

where $k \in \mathbf{Z}$, $k \geq 0$, and

(a) A runs over all smooth one-forms on S^3 with values in $su(2)$, modulo the gauge group of all smooth maps from S^3 to $SU(2)$.

(b) "tr" is some appropriate normalization of the natural trilinear form on $su(2)$.

(c) L is a link with components C_j.

(d) Given a component C_j and an A, $P \exp\left(\int_{C_j} A\right)$ is the trace of parallel transport along C_j for a 2-dimensional fibre.

Although the measure $(\mathcal{D}A)$ has not yet been shown to exist, it is possible to examine the formal consequences of its existence in terms of Witten's topological quantum field theory, in which one may cut up a 3-manifold piece with surface boundary, and the theory asserts the functorial existence of a vector space for each surface and a vector in this space whenever the surface bounds a 3-manifold. By a heuristic identification of the surface with the conformal blocks of the WZNW theory referred to in the previous section, Witten was able to show that the functional integral satisfied the recursion relation satisfied by V, and thus identify it with $V_L\left(e^{\frac{2\pi i}{k+2}}\right)$.

The great advantage of Witten's theory is that S^3 plays no special rôle and the theory works for any closed oriented 3-manifold. Moreover, the topological quantum field theory fits extremely well into the surgery framework, so Witten was able to give explicit formulae for his invariants of links (possibly empty!) in an arbitrary 3-manifold. These formulae have been checked via the Kirby calculus in its Fenn–Rourke version.

Algebraic Quantum Field Theory

In algebraic quantum field theory, one supposes a Hilbert space and, to every reasonable region O of space-time, a von Neumann algebra $\mathcal{A}(O)$ of observables localised in that region. The algebras are supposed to obey certain natural relations such as $\mathcal{A}(O) \subseteq \mathcal{A}(O')$ if $O \subseteq O'$. The most important such relations are *covariance*, which gives a unitary representation of the Poincaré group on the Hilbert space moving the $\mathcal{A}(O)$ around in the obvious way; *positivity*, which is a statement about the spectrum of the representation of the Poincaré group, basically asserting that energy is positive; and *causality*, which asserts that $\mathcal{A}(O)$ and $\mathcal{A}(O')$ commute if O and O' are space-like separated regions, i.e., no particle travelling at less than or equal to the speed of light can go from O to O'. With these axioms, it is known that $\mathcal{A}(O)$ is always in a hyperfinite III factor whose uniqueness was shown by Connes and Haagerup. Although it is hard to argue that the axioms should not be satisfied in quantum field theory, one would not expect to be able to deduce much from them, since there is no dynamic content, such as a Lagrangian.

Thus it was remarkable that Dopplicher, Haag and Roberts found a lot of structure when looking at "superselection sectors." The idea is that a theory may decompose into sectors (a direct sum decomposition of the Hilbert space) corresponding to some conserved quantum number (e.g. charge). There would then be a privileged "vacuum" sector where that charge is zero. Their theory turns around two ideas: The first is that all the sectors should be equivalent to the vacuum sector away from some bounded region, and the second is that, in the vacuum sector, "Haag duality" should hold, i.e., the commutant $\mathcal{A}(O)'$ of $\mathcal{A}(O)$ is $\mathcal{A}(O')$ whenever O is the intersection between a forward and a backward light cone and O' is the set of all points space-like separated from O.

Considering this structure, Dopplicher, Haag and Roberts were able to deduce the existence of an *endomorphism* $\rho : \mathcal{A}(O)' \to \mathcal{A}(O')$, and of representations of the symmetric group coming from ρ and the consequences of moving regions around $\mathbf{R}^4$.

It has been observed by many people (Fredenhagen–Rehren–Schroer, Fröhlich et al, Longo) that, if one applies the Dopplicher–Haag–Roberts theory in dimension less than 4, the symmetric group is replaced by the braid group. It is expected that if one could construct the quantum field theory of the WZNW model in such a way as to be able to analyze the algebras $\mathcal{A}(O)$, the braid group representations would be those calculated by Tsuchiya and Kanie. Some progress has been obtained in this direction by Wasseman.

But notice that the algebraic theory never uses conformal invariance, so one would expect braid group representations in any low-dimensional quantum field theory. It would be interesting to try to identify the parameters of the braid group representations in non-conformal field theories, e.g. the one developed by Zamolodchikov in his E_8 perturbation of the Ising model.

Finally, it seems that a complete analysis of these four theories and their connections will take mathematicians and physicists many years yet, but note that, so far, these ideas have not helped to answer the two elementary questions posed earlier about the knot polynomial $V_L(t)$.

Vaughan Jones
Mathematics Department
University of California
Berkeley, California 94720
USA

Transcribed from the videotape of the talk by Carme Safont; revised and supplemented by the author.

Symposium on the Current State
and Prospects of Mathematics

Barcelona, June 1991

Recent Progress in Diophantine Geometry

by

Gerd Faltings

Fields Medal 1986

for his proof of the Mordell, Shafarevich and Tate conjectures using methods of arithmetic algebraic geometry to compute the height of the Abelian varieties considered as points in a suitable space.

Abstract: Diophantine problems have a long history. I review recent results, due to P. Vojta, G. Frey and myself. In the first half of this century, Thue and Siegel developed the method of Diophantine approximation to prove finiteness results for the number of rational or integral points on certain curves. However, they did not manage to prove the Mordell conjecture, which in some sense is the strongest possible statement of this kind. This has been achieved only recently by P. Vojta, and his methods allow generalisations which settle the question for arbitrary subvarieties of Abelian varieties. However, before that, Arakelov and Parshin had developed other methods which work for function fields, and I myself had managed to prove the conjecture with them. Also recently Masser and Wüstholz have used methods from transcendence theory to give effective versions of some results.

I intend to review these and also explain on the way some nice mathematics which has come out of this, namely Arakelov theory. Finally, I intend to cover the relation between Fermat's conjecture and elliptic curves, which was discovered by G. Frey.

Recent Progress in Diophantine Geometry

This is a report on some new results obtained during the last ten years. Naturally, I can only sketch proofs and have to leave out many interesting topics. I thus only discuss things I feel competent about, and this should not be seen as any negative judgement about others.

Roughly, after a very general introduction, I first report on progress in Diophantine approximation, due to P. Vojta and myself. This method has been known since the beginning of this century, but only recently has it been able to produce a proof of the Mordell conjecture. For this one had to introduce various tools from algebraic geometry. After that, I shall report on the Parshin–Arakelov method, by which the Mordell conjecture was first shown. It also yields the Tate conjecture, something apparently not possible using Diophantine approximation. However, there is also now a new proof for this, using Baker's method (Masser–Wüstholz). In fact, it gives an "effective" version. So we have now two approaches to most results: one using classical methods and one more geometric using Arakelov-style geometry (which, however, was the first one to produce proofs). I take the opportunity to give a short description of Arakelov theory, a method by which one tries to carry over ideas from the case of function fields to number fields. Finally, I describe G. Frey's idea by which he relates Fermat's last theorem to elliptic curves. However, this is a sort of exception, as generally I try to steer clear of modular forms.

Heights and Finiteness

The object of Diophantine geometry is to look for the integral solutions of equations. Given finitely many polynomials $F_1, \ldots, F_r \in \mathbb{Z}[X_1, \ldots, X_d]$, we want to find d-tuples of integers, $x_1, \ldots, x_d \in \mathbb{Z}$, such that $F_i(x_1, \ldots, x_d) = 0$ for all i. This may be reformulated as follows: The zero set of these polynomials is an affine variety X in the affine space,

$$V(F_1, \ldots, F_r) = X \subseteq \mathbb{A}^d,$$

and we are interested in the integral points of X. For example, in two variables, one single equation defines a real curve and we ask for the points of integer coordinates in the plane belonging to the curve.

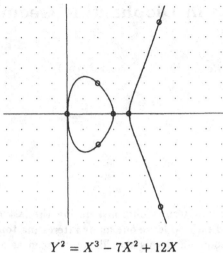

$$Y^2 = X^3 - 7X^2 + 12X$$

One would like to have algorithms to find all integral points, but this is very difficult; so, in general, we have to talk about *finiteness* results.

A general procedure to prove that there are finitely many solutions is to *bound the height* of the solutions. The height of a point is just the length

$$H(x) = \|x\| = \sqrt{x_1^2 + \cdots + x_d^2} \, .$$

If the height of the integral points of X is bounded, then all points are contained in a (big) ball and there are only finitely many integral points inside this ball. This is called an *effective estimate*. Unfortunately, in real life it is rarely possible to do this, so things are more complicated.

Nevertheless, it is very important to have good height functions for various problems. For a rational number $x = p/q$, $\gcd(p, q) = 1$, one defines its *(big) height* $H(x) = \sqrt{p^2 + q^2}$ and its *(small) height* $h(x) = \log H(x)$. This notion extends to projective spaces and, more generally, to each pair consisting of an arithmetic variety together with a line bundle on it. If the line bundle is ample, then any set of rational points of bounded height is finite. Usually, one has some moduli space (for curves, or Abelian varieties, or a Hilbert scheme, or a Chow variety), so one has to find an ample line bundle on it and a good way to compute heights, not involving too much of the complicated structure of the space. For example, for Abelian varieties one takes as line bundle ω, the bundle of modular forms or better the top-degree differentials on the Abelian variety. Then one can compute the height $h(A)$ of an Abelian variety solely from A itself, without using projective embeddings of the moduli space. However, as this space is noncompact, one still has to solve some problems about the boundary. But once this is done one can forget about them. Similarly, one can also define heights for cycles in projective space. Here the moduli space is the Chow variety, which is compact but about which little else is known (like its dimension). Nevertheless, one can define a good height function which works nicely.

I go back now to our original setting. For obvious reasons, it is very natural and convenient to work with projective varieties. This means that one adds some points at infinity by adding a new variable X_0 and converting all F_i into homogeneous polynomials. Two solutions differing by a scalar multiple are identified. In particular, for projective varieties there is no more difference between rational and integral points. Thus, I will mostly talk about rational points. There are usually three classes of projective varieties, which I may call *elliptic, parabolic* and *hyperbolic*. One may talk about *positive curvature, curvature zero* and *negative curvature*. This has a similar meaning to the usual one but not quite the same. Roughly speaking one has:

$$\text{hyperbolic} \implies \text{finiteness of rational points}.$$

A curve of genus greater than one is an example. In the function field case this can be seen very easily. If we have a curve C over a base curve B (think for instance that C is a constant curve), a rational point is a mapping $\varphi : B \longrightarrow C$ and it has been known since the old days that there are only finitely many such mappings. The proof is the following: From Hurwitz's formula,

$$2g(B) - 2 = (2g(C) - 2) \cdot \deg \varphi + \text{a nonnegative term,}$$

we see, since the genus of C is at least two, that the degree of such a mapping is bounded. After some arguments, one can see that the degree corresponds to the height of the point and finiteness is obtained.

Abelian varieties are the standard objects belonging to the parabolic case. They can be described as projective, smooth, connected algebraic groups. The Jacobian variety of a curve is an example of Abelian variety. Some complex tori are also examples of Abelian varieties over the complex numbers. A complex torus is a complex vector space of some dimension divided by a lattice of rank twice the dimension; this object may not be algebraic at all, but those complex tori satisfying the so-called *Riemann relations* are Abelian varieties.

For Abelian varieties we do not have finiteness of the number of rational points, but still there is some sort of finiteness statement: the theorem of Mordell–Weil. This theorem asserts that for any Abelian variety defined over $\mathbf{Q}$ (or over any number field), the Abelian group $A(\mathbf{Q})$ is finitely generated. How does one prove such a thing? One takes an ample line bundle L on A, which gives an embedding $A \hookrightarrow \mathbf{P}^N$ into some projective space. From this embedding we get a height function on $A(\mathbf{Q})$, just by taking the usual (little) height on $\mathbf{P}^N(\mathbf{Q})$. This height is a positive definite quadratic function on $A(\mathbf{Q})$. One also shows that $A(\mathbf{Q})/2A(\mathbf{Q})$ is finite, which follows from some Galois computations. From both facts it is easy to show that the rational points up to some bound on the height must generate the group.

This was proved in the twenties. By that time, the conjecture of Mordell had already been raised, predicting that curves with genus greater than one have only a finite number of rational points. Since the curve is lying in its Jacobian and the rational points of the Jacobian form a finitely generated group, it was very natural to try to prove that the curve intersected with a finitely generated group could contain only a finite number of points. Many people have tried to use this idea, but nobody except (recently) Vojta has been able to use this finite-generatedness to prove finiteness. I will come to this in a moment, but first I will go back to history again, namely to finiteness theorems obtained with techniques of Diophantine approximation.

Diophantine Approximation

The method of Diophantine approximation usually works as follows, demonstrated by the example of Roth's theorem. Suppose α is an algebraic number. We want to show that, for any exponent $2 + \varepsilon > 2$, there are only finitely many rationals $x = p/q$ for which $|x - \alpha| \leq q^{-(2+\varepsilon)}$. To do this, we assume that there are infinitely many $x_i = p_i/q_i$. Choose them such that q_1 and the ratios $q_2/q_1, q_3/q_2, \ldots, q_n/q_{n-1}$ are all big ($n > 0$ sufficiently big). Then construct a polynomial $f(t_1, \ldots, t_n)$ which has a high-order zero at $(\alpha, \ldots, \alpha)$. It then follows that f also vanishes to high order at $(x_1, \ldots, x_n)$, as for example one can bound the denominator of its value, and estimate its absolute size because the x_i are close to α. On the other hand, one has to derive that this cannot happen, which we will do below using the product theorem. For the construction of f one has to count dimensions; instead of considering polynomials over $\mathbf{Q}$, one can also consider polynomials over $\mathbf{C}$ which vanish at all conjugates of $(\alpha, \ldots, \alpha)$. One shows that the number of conditions imposed by one such high-order zero decreases very fast provided the order is less than a half of the degree, and n is big. Thus one can use Siegel's lemma to construct f. Also the exponent $2 + \varepsilon$ derives from this construction, i.e., $1/(2 + \varepsilon)$ is less than $1/2$.

Now let us explain the product theorem. The key to it is that $f(x_1, \ldots, x_n)$ should have x_i-degree d_i such that $d_i \log(q_i)$ is approximately constant, so that $d_1 >> d_2 >> \cdots >> d_n$. Also the order of zero of f should be about $d_i/2$ in x_i-direction. Now consider for example the case $n = 2$, and $Z_\sigma \leq \mathbf{P}^1 \times \mathbf{P}^1$ denote the locus where all derivatives $\partial_1^{i_1} \partial_2^{i_2}(f)$ vanish for $i_1/d_1 + i_2/d_2 \leq \sigma$. Fix a small $\varepsilon > 0$ and let Z denote a common irreducible component of Z_σ and $Z_{\sigma-\varepsilon}$. We claim that Z is a product, $Z = Z_1 \times Z_2$, provided d_1/d_2 is big enough.

To see this, assume that Z is a curve such that both projections to $\mathbf{P}^1$ are dominant. Then Z occurs with multiplicity $\geq \varepsilon d_1$ in a curve defined by an equation of bidegree (d_1, d_2), i.e., $\varepsilon d_1 Z \leq d_1 H_1 + d_2 H_2$ (the H_i are pull-backs of hyperplanes on $\mathbf{P}^1$). Multiplying with H_1 we obtain $\varepsilon d_1 (Z H_1) \leq d_2$. Thus $Z H_1 = 0$ if $\varepsilon d_1 > d_2$, and the projection from Z to the first factor is constant. This contradiction shows the assertion.

Furthermore, from the proof one can estimate the heights of the Z_i. One obtains that $d_1 h(Z_1) + d_2 h(Z_2)$ is bounded by $(d_1 + d_2) h(f)$. Now, in our case, either Z_1 or Z_2 must be a divisor containing x_1 respectively x_2, so $d_1 h(x_1) + d_2 h(x_2)$ can be bounded. This, however, is a contradiction if $h(x_1) \approx \log(q_1)$ is chosen big enough.

So, in short, we first choose n big enough so that we can construct f, and then q_1 big enough, $\log(q_2)/\log(q_1)$ big, etc., so that we can apply the product theorem, and the above argument gives the required contradiction. One also notes that the method of proof is indirect; one does not obtain an explicit bound for the q_i's. This is a common feature of all such proofs.

In dimension one (like Roth's theorem) we only need to apply the product theorem once, as all proper subvarieties are points. For higher dimensions, one has to use an inductive procedure: By the product theorem, $(x_1, \ldots, x_n)$ is contained in a product $X_1 \times \cdots \times X_n$, where at least one of the X_i is different from the whole space, and one can bound the heights of the factors. One then starts again by constructing a new f on this product, which in turn leads to a smaller product, etc. For example one can derive W. Schmidt's theorem in this manner.

For Abelian varieties, one can considerably strengthen these results, using a wonderful idea of P. Vojta. It makes use of the Mordell–Weil theorem. Although known for more than sixty years, before Vojta nobody could really apply it to the Mordell conjecture. What Vojta does is to observe that on products of Abelian varieties there are more line bundles than just pull-backs from factors, namely also Poincaré bundles. They allow "custom-made" bundles to be produced which work particularly well for a given rational point, and now f takes values in such a bundle. Using this we obtain a Roth-type result for any exponent > 0, instead of > 2. Also for the case of subvarieties $Z \subseteq A$ which do not contain any translate of a positive dimensional Abelian subvariety, one obtains finiteness of rational points for such a Z. However the term "Diophantine approximation" is somehow misleading in this case, as there is nothing to approximate. An example of such Z's is obtained for curves C of genus $g \geq 4$ which are not a double cover of $\mathbf{P}^1$ or an elliptic curve: For them, the second symmetric power injects into the Jacobian $J(C)$, and its image contains no translate of an elliptic curve (a result due to Harris and Silverman). This example is also of some interest as its geometric fundamental group is Abelian, which is not the case for the other types of varieties for which finiteness was known before.

The Parshin–Arakelov Method

This method reduces the Mordell conjecture to the Shafarevich conjecture, which states that there are only finitely many curves (or Abelian varieties) of a given genus (respectively of a given dimension) which have good reduction outside a given finite set of places of a number field. The reason why this is possible is given by a construction of Kodaira and Parshin: To a point x on a curve C one associates a new curve $C(x)$ which is a Galois covering of C ramified precisely in x. This is possible as $\pi_1(C \setminus \{x\}) \neq \pi_1(C)$ (however, the covering is non-Abelian). This gives a map from C to the moduli space $\mathcal{N}_g$ of curves of genus g (where $g \gg \text{genus}\,(C)$), or to the moduli space $\mathcal{A}_g$ of principally polarized Abelian varieties of dimension g. In the case of function fields, one shows that curves with given bad reduction have bounded height, and admit no deformation provided they are not isotrivial, i.e., they are not twisted versions of constant curves. For the first statement, one uses Hodge theory, which relates coherent cohomology to singular cohomology. The dimension of the latter is easily bounded using the Leray spectral sequence. For the deformation result one needs the Kodaira–Spencer class to be nontrivial, which implies some positivity for the relative differentials, and then one uses Kodaira's vanishing theorem. The first step also works for Abelian varieties, and suffices to show the Mordell conjecture. The deformation theory, in turn, does not work in this case.

However, all this makes heavy use of the differential calculus over the base field, and thus cannot be generalized to number fields. To treat these, one needs a new idea, using the associated ℓ-adic Galois representations: One shows that they are semisimple and determine an Abelian variety up to isogeny. This is the Tate conjecture, which had to be shown on the way. Then one uses the fact that the trace of any Frobenius element is an integer of bounded size (by Weil's theorem), so for a fixed Frobenius there are only finitely many choices. By a variant of Minkowski's theorem, finitely many such traces determine the representation, and the assertion follows, that is, one already has finiteness up to isogeny. Finally, one shows that any isogeny class is also finite, by modifying the arguments in such a way that they work mod ℓ, for a sufficiently big prime ℓ.

Baker's Method

This procedure has the advantage that it yields effective estimates, i.e., one can —at least in principle— bound the size of solutions. However, it does only apply to certain problems about algebraic groups. Recently it has been used by D. Masser and G. Wüstholz to give an effective version of the Tate conjecture, that is, one can bound the minimal degree of an isogeny between two Abelian varieties, provided one knows that such an isogeny exists. To do this, one uses zero-estimates to show that if an Abelian variety has Abelian subvarieties, then it has such of bounded degrees. Applying this to a product yields the result. However, this method does not give an effective bound for the Shafarevich conjecture. We thus have now two independent proofs for the Mordell as well as the Tate conjecture.

Arakelov Theory

The integers are in some sense analogous to an open affine curve. It is well-known that affine curves are not good because of their lack of nice global invariants. If one wants to get global invariants, one needs projective (complete) curves. An affine curve can be completed by looking at all valuations of its function field and one can try to do a similar thing in the integers. There, one has the usual p-adic valuations which allow one to pass from $\mathbf{Q}$ to $\mathbf{Z}$, but the infinite valuation of the rationals is also needed to get a good picture. Of course, this is a very different valuation since it is Archimedean. But it is possible to treat all valuations in the same manner, by translating everything in terms of metrics. This point of view has been known for a long time in dimension zero (analogy between number fields and function fields over finite fields), yet, in higher-dimensional geometry, the first consistent theory is due to Arakelov.

Arakelov theory translates methods from algebraic geometry into analysis. The main principle is: If we have to use algebraic geometry over the integers, then every object has to be endowed with a metric. For example, a $\mathbf{Z}$-module M can be described by its associated $\mathbf{Q}$-vector space together with a family of p-adic metrics, one for each prime number p.

$$M \iff (V = M \otimes_{\mathbf{Z}} \mathbf{Q}, \{\| \ \|_p\}).$$

M can be recovered as the set of points of V of length less than or equal to one with respect to all p-adic metrics. The seek of compactification in analogy with geometry suggests to add a metric at infinity: a positive definite quadratic form on $V \otimes_{\mathbf{Q}} \mathbf{R}$ or a Hermitian form on $V \otimes_{\mathbf{Q}} \mathbf{C}$. As an example, let us think what should be a line bundle on the compactified line. A line bundle over $\mathbf{Z}$ is just a projective $\mathbf{Z}$-module of rank one; since they are all free, there are no interesting isomorphy classes. But if such a module is endowed with a metric, then one gets a more interesting theory. Giving a metric is essentially the same as giving $\|1\|$, the norm of one. Thinking in this line bundle as a variety, there is a sort of degree:

$$degree = -\log \|1\|,$$

which is an invariant of the module.

Similarly, to endow a scheme X over $\mathbf{Z}$ with a structure at infinity, one specifies a Kähler metric on $X_{\mathbf{C}}$, stable under complex conjugation. For this, one needs $X_{\mathbf{Q}}$ to be smooth, a general feature of the theory due to the fact that we do not know how to do analysis on singular spaces. Also a vector bundle E on X should get a Hermitian metric over $X_{\mathbf{C}}$. Then a sequence of bundles $0 \to E_1 \to E_2 \to E_3 \to 0$ is exact at infinity only if it is split in the metric sense. In general this is not the case, and one obtains a secondary class which describes the cohomology at the infinite fibre. An analog at finite places is given by an exact sequence of bundles over $\mathbf{Q}_p$, so that the cohomology is p-torsion. Using this, Gillet and Soulé have constructed an arithmetic K-theory and also arithmetic Chow groups. Furthermore, together with Bismut, they have recently shown an arithmetic Riemann–Roch theorem; here the direct image of a bundle E is represented by the alternating sum of the C^∞ sections of $E \otimes \Omega^{0,q}$. To make sense out of it, one needs a regularization procedure using heat kernels. All in all, Arakelov theory is a beautiful mixture of algebra and analysis. However, one ingredient still missing is the analog of Hodge theory and singular cohomology. So far, there are only vague ideas about "motivic cohomology," with the conjectures far outnumbering the theorems.

Fermat and Elliptic Curves

Recently, G. Frey has found a connection between Fermat's conjecture and elliptic curves, which so far seems to be the most compelling reason why the conjecture should be true. In short, if $a^p + b^p = c^p$ (p prime, a, b, c integers), consider the curve given by $y^2 = x(x - a^p)(x - c^p)$. Its discriminant is $4a^p b^p c^p$ and is much bigger than its conductor $2abc$. If we had a good effective Mordell theorem, we could derive that this cannot happen for big primes p. Also, if the curve were a Weil curve, we could find a modular form of weight two associated to its L-series. Ribet has shown that, modulo p, one can decrease the level of this form until it is one. However, there are no such forms. Thus Fermat's conjecture should follow if every elliptic curve over $\mathbf{Q}$ were a Weil curve, which is widely believed to be true, yet has not been demonstrated so far.

References

[1] S. Arakelov, Families of algebraic curves with fixed degeneracies, *Izv. Akad. Nauk SSSR Ser. Mat.* **35** (1971); English transl. in *Math. USSR Izv.* **5** (1971), 1277–1302.

[2] S. Arakelov, Intersection theory of divisors on an arithmetic surface, *Izv. Akad. Nauk SSSR Ser. Mat.* **38** (1974); English transl. in *Math. USSR Izv.* **8** (1974), 1167–1180.

[3] J. M. Bismut and G. Lebeau, Complex immersions and Quillen metric, preprint Orsay (1990); also *C. R. Acad. Sci. Paris Sér. I Math.* **309** (1989), 487–491.

[4] P. Deligne, Le déterminant de la cohomologie, *Contemp. Math.* **67** (1987), 93–177.

[5] G. Faltings, Calculus on arithmetic surfaces, *Ann. of Math. (2)* **119** (1984), 387–424.

[6] G. Faltings, Endlichkeits-sätze für abelsche Varietäten über Zahlkörpern, *Invent. Math.* **73** (1983), 349–366.

[7] G. Faltings, Diophantine approximation on Abelian varieties, to appear in *Ann. of Math. (2)*.

[8] H. Gillet and C. Soulé, Arithmetic intersection theory, *Inst. Hautes Études Sci. Publ. Math.* **72** (1990), 93–174; also *C. R. Acad. Sci. Paris Sér. I Math.* **299** (1984), 563–566.

[9] H. Gillet and C. Soulé, Characteristic classes for algebraic vector bundles with Hermitian metrics, *Ann. of Math. (2)* **131** (1990), 163–203, 205–238.

[10] H. Gillet and C. Soulé, Analytic torsion and the arithmetic Todd genus, *Topology* **30** (1991), 21–54.

[11] H. Gillet and C. Soulé, An arithmetic Riemann–Roch theorem, *C. R. Acad. Sci. Paris Sér. I Math.* **309** (1989), 929–932.

[12] S. Lang, *Introduction to Arakelov Theory*, Springer-Verlag, New York, 1988.

Gerd Faltings
Department of Mathematics
Princeton University
Fine Hall, Washington Road
Princeton, New Jersey 08544-1000
USA

Text by the author, supplemented by Enric Nart with material from the videotape of the talk.

Symposium on the Current State
and Prospects of Mathematics

Barcelona, June 1991

Round-Table Discussion

Invited Participants: Alain Connes
                                 Gerd Faltings
                                 Vaughan Jones
                                 Stephen Smale
                                 René Thom

Moderator: Jorge Wagensberg

At the end of the Symposium, a round-table discussion was organized jointly with the Science Museum of Barcelona. The moderator was Professor Jorge Wagensberg, who is a member of the Faculty of Physics of the University of Barcelona and current Director of the Science Museum.

The purpose of the round table was to highlight in a common framework several relevant aspects of present-day mathematical thinking —most of which had already been touched on within the preceding talks— and the relationship of mathematics with other sciences and technologies in the present as well as in the future. This was very much in accordance with the spirit of the Symposium, which achieved in this final session the necessary summary of conclusions.

Several themes had been proposed to the invited participants, who met together with the organizers prior to the round table in order to define the main directions of the discussion. The proposed subjects included methodology of mathematical creation in comparison with other sciences; rôle of computers in modern and future mathematics; relationship of mathematical modeling with reality; new trends in the production and diffusion of mathematics. The audience was encouraged to join the discussion after the invited participants completed a preliminary debate on these key themes.

Round-Table Discussion

Wagensberg: Ladies and gentlemen, welcome to the last session of this very interesting meeting. I do not need to introduce again our invited professors, so I simply mention their names: Professor Thom, Professor Connes, Professor Smale, Professor Jones, Professor Faltings. The purpose of this round-table discussion is to make some comments on several general and concrete questions on mathematics. Our plan is to start with questions to our invited professors and, once the temperature of the audience has reached some critical value, we shall open the debate for the audience.

I would begin with a rather general question. Science is, I think, what scientists say that science is. But there are some criteria, since scientists have referees who accept or reject the papers in the journals. The referees have some criteria for distinguishing between what is science and what is not. We could also start with the same idea in mathematics: Mathematics is what mathematicians say that mathematics is. But the question would be: What are the criteria in mathematics? That is, what is the difference between the kind of knowledge called mathematics and other kinds of knowledge like science, art, etc? I open the debate.

Connes: I think this is a fairly general question. I just want to point out that it is quite easy for a scientist, like a chemist or an astronomer, to explain to the general public what the object of his science is. In chemistry, or physics, or biology, or astronomy, the object is, in some sense, the study of matter at different scales. Mathematics has two aspects which make it difficult to explain, and which make it different from other sciences. I think the first aspect is that mathematics is, in many ways, used as a language in other sciences. It is tempting for other scientists to try to reduce mathematics to a language, in the sense that, if they use mathematics —in a model or something like that— they do not use mathematics as we do. They use it as a language. It is quite difficult to explain that mathematicians are also studying an aspect of reality which is not material (i.e., which has nothing to do with matter), and it is also difficult to try to make precise what one means by *mathematical reality*. These two aspects, mathematics as a language, and mathematics as the study of some kind of reality (which is difficult to define), make it different from other sciences.

Smale: More than most mathematicians, I tend to think that mathematics is more like art than other sciences. But there is one special difference, I find, which is that mathematics tends to be correct. Mistakes in mathematics are rarely significant for very long. But also mathematics tends to be more irrelevant. There is so much of mathematics that tends to go off in directions which are appreciated only by a very few, irrelevant even to the rest of mathematics. So I think there is a bigger danger in mathematics than there is in other sciences of tendencies to go off into irrelevancies, i.e., into things that are correct but not important.

Faltings: The main difference between mathematics and physics, for example, is that physics really lives from the experiment. If you have a brilliant theory, but the experiment does not confirm it, then you are lost. Mathematics is only required to be correct, interesting, and useful. I think this is one difference between us and other scientists.

Jones: My view is that there is a kind of continuum, so that we can do different kinds of mathematics, from the purer to the more applied, where it gets closer to physics. It is all science as far as I am concerned.

Thom: I think it was the French writer Paul Valéry who said that when you say that a question is *philosophical*, you can abstain without any trouble from considering it. And the problem of the relation between mathematics and reality is fundamentally a philosophical problem. Hence, if you believe Paul Valéry, you could certainly abstain from thinking about it. But if you do not share this very immediate and pragmatic point of view, then you have to consider the problem of the relation between mathematics and reality as one of the basic problems of philosophy, perhaps even of metaphysics. And I do not think one may be completely indifferent to this question. After all, one of the oldest philosophical problems deals with the distinction between Plato and Aristotle, and this is deeply associated with the question of the relation of mathematics with reality. Let me just say a few words about my general feeling about that. If you are dealing —as Professor Connes was saying earlier— with scientists who claim that mathematics is nothing but a tautology dealing with our neuronal activities, then you may ask these scientists: What is the status of space? Is space something existing outside of us, or is it really just an imagination of our mind? I suspect the only reasonable answer is that space exists. If you say that space is nothing but the play of our neurons, as many neurophysiologists would currently say, then you might put them the following question: If you do not believe in the existence of this bottle outside of me, how would you believe that I could accept the existence of neurons and synapses?

As a result, one cannot avoid considering the problem of existence of space as, so to speak, the midpoint where subjectivity and objectivity meet together. My own philosophy is that subjectivity and objectivity meet on the level of existence of space —space understood subjectively as a continuum, and objectively as the diverse theoretization that mathematics and physics have developed about space. But, fundamentally, space and time are the primitive experience, and it is on this stage that the drama takes place.

Wagensberg: This suggests the following question. In physics, for example, it very often happens that a theory, like relativity or quantum mechanics, gets extended to philosophical thought, more or less directly. What is your opinion about the same thing in mathematics? For example, Gödel's incompleteness theorem is a very clear example where some philosophical thought is exported from mathematics to the outside world of thought.

Connes: I think Gödel's theorem can be taken as a good example, because, depending on the way one looks at it, one can make it serve several different purposes. For instance, some people criticize mathematics using Gödel's theorem, by saying "Look, you cannot even prove that you are doing something which is not contradictory." And it can be used in different ways. So it is very difficult to assess whether the way it is used in philosophy is sound.

One way to look at it, which I consider to be relevant, is the following (this way is in fact very little known among mathematicians themselves). If we look at the way mathematics has been formalized by Bourbaki, there Gödel's theorem is considered, essentially, as saying that, in a given system of axioms, there will always be some mathematical problem which will be undecidable. But, in fact, the theorem says much more. The theorem says that if we consider any reasonable system (i.e., any formal language), which we assume to be noncontradictory, then there is a certain proposition about integers which is true, but which is not decidable in the given system. One way to interpret this is the following: There is this sort of fairly primitive mathematical reality of the integers, but it is impossible to comprehend all its properties by means of the sort of projective axiomatic systems that we construct in trying to comprehend it. So it makes rather clear the opposition between something which is a sort of primitive mathematical reality (which is what I was alluding to before) and the systems that we construct with tools in order to try to understand this reality.

This is just to give an example of how Gödel's theorem can be used in several directions. So I have my own reservations on the use of this in pure philosophy, because I think it is quite dangerous.

Wagensberg: Perhaps on the level of ideas, rather than on the level of results.

Connes: Yes. I think it is important to have a debate on it, and to have contradictory opinions, rather than just one opinion.

Smale: I would think that the biggest danger with the Gödel theorem is that, while having some philosophical implications, it is misused by philosophy. A good example, I would say, is in Penrose's new book. Many of you have read *The Emperor's New Mind*, where he takes Gödel's theorem as a big argument against artificial intelligence. I think there is a big fallacy in that, because he is arguing against continuous models —a continuous kind of mathematics in which both physics and the brain work— and he is using a highly discretized version of Gödel's theorem, which is already highly discrete itself. He uses Turing machines, which I think have very little to say about things where the real numbers are involved. It is much more useful to take into account Tarski's theory, where it is shown that the reals are decidable, thus providing a decidability result for the real numbers which is more applicable to physics and biology than the Gödel theorem.

Wagensberg: There is a paper of Lucas on this subject. He used the Gödel theorem in order to prove that minds are not machines.

Smale: It sounds a little bit like Penrose's argument. I would probably have the same objection to Lucas' as I do to Penrose's.

Jones: I am really not worried too much about these things. I have always felt that, if one day someone came up with a contradiction in mathematics, I would just say "Well, those crazy logicians are at it again," and go about my business as I was going the day before. It would not worry me too much.

Thom: Our ancestors first believed that the Earth was flat. Later, with some thinking, they found that the surface of the Earth was round, spherical; and later on decided that space had to be Euclidean. Some centuries later, we decided that space-time had to be Minkowskian. And now, with general relativity, we have a lot of models, among which we do not know exactly how to choose. So, I suspect that all the theoretical constructions one can make about the global structure of the universe will always be dependent on local extrapolating mechanisms which will be mathematical, essentially extrapolations of symmetry. We have local symmetries around us (essentially, let us say, classical Euclidean geometry), which have at least the advantage of having some solid foundation in our body constitution. Each time we act with our hands, each time we compute the motion of a solid body, we do things which are in some sense ingrained in our own biological constitution. In that respect, this is a much stronger basis than all the extrapolation of physicists.

Wagensberg: Let me raise a second question that is related to our discussion now. I think that we are currently living through a significant moment regarding computers. Computers are changing very fast. And the question is: What is their impact on mathematics? I think there are a lot of questions that are related to this one. For instance, are computers only a tool? Can they help mathematical creativity in mathematical production, as is the case for design in architecture? (In this case, they are more than a tool.) What is the credibility of a computer result?

In physics we now have simulation, which is different from calculation. We have had two hundred years of experiments, and perhaps ten years with simulation. It is not the same thing if we find some experiment contradicting relativity theory, or a computer simulation contradicting relativity theory. So what is the merit, for example, of a proof obtained by a computer?

Another interesting question may be how to use computers in the teaching of mathematics.

Smale: I have some strong feelings about that. I think that computers help a lot in the day-to-day practice of experimental mathematics, but there is a much deeper effect of computers on mathematics, which is changing the whole structure of mathematics. Many of you heard me say recently that the greatest problem in mathematics in the last half of the twentieth century comes from the computer. It is the problem "Is $NP \neq P$?" It is creating a whole different way of thinking about every subject in mathematics. It is about the way we do mathematics, about the emphasis on the algorithms. And to me that is going to be the long-run deep impact of the computer in mathematics. All the other ways I think are important, but this one transcends all that for mathematics itself.

Jones: I use computers quite a lot. I have several, perhaps contradictory, attitudes towards them. One thing I do not like is computer-aided proofs. For instance, this four colour theorem proof leaves me very unhappy. It is conceivable that the computer actually made a mistake, and would repeat it no matter how many times we ran that program.

Maybe there is an electrical fault, or something. One cannot really believe it, and one does not really understand it. On the other hand, I do a lot of calculations of polynomials of knots, and there I would never trust a hand calculation. If it is not done by a computer, I just would not believe it. I think that is an interesting attitude.

In fact, my main use of computers at the moment is in simulation, and I think this is going to raise a lot of interesting issues in science, perhaps not so much in pure mathematics. But the problem I am looking at is a topological problem to do with DNA. DNA gets tied up in links, and it has to get itself untied. The question is how it is done. So far this is not an observable phenomenon. We can sort of control the input and have a look at what comes out of an experiment, but we cannot actually see what is happening while this untying is going on. So, in some sense, before the advent of computers, there was nothing that could be done about trying to see what was going on in the meantime. But now, what I am actually doing is to run a simulation of this process. It is proposed that there is some kind of stochastic unlinking going on, so there are some enzymes, whatever they are, closing crossings in the two DNA molecules, and then there is some force pulling the two molecules apart. And the question is: Is that a sufficiently good mechanism to explain the speed at which the DNA molecules do indeed come apart every time a cell divides? Well, the computer is wonderful for that. We could not imagine doing any kind of analytic estimate of the time that it would take for these two DNA molecules to come apart. But the program is running now. We can sort of see these two molecules, and watch them as they come apart, and see how everything is going on. And we can test various hypotheses. The interesting thing is that we have written a program that is just the vaguest of guesses as to what might be going on. So how can we be sure that it has anything to do with what is actually happening in the cell? Well, I do not know. How do we know that our computer program is working right? This is very important. One of the main things here is actually to be able to see what is happening. So, in fact, a very important part of this computer project is to have good graphics, so we can control, and see that everything is not going wild. Thus, I think there are some very interesting problems coming up from simulation, as to whether what the computer is doing actually has anything to do with the realities.

There are similar problems in solving some complex differential equations, even the Navier–Stokes equation near turbulence. How do we know that we are actually solving anything to do with turbulence? How do we know it is not just an artifact of the computer? Once again, it is very important to have a visual picture of what is going on. I think these are interesting scientific questions coming out of the computers, and computers will have a big impact.

Wagensberg: I think that this raises a methodological question. In many conferences, in many scientific debates —at least in physics— the theme of discussion is "What is the value of a result obtained by computer simulation?" Many times this is offered as a substitute for theory, and at other times as a substitute for experiments. And perhaps we are now in our third approach to reality besides theory and experiments.

Jones: That is right. I think this is really the problem. What can we actually conclude from the simulation? We are not really sure that it corresponds to all these computer solutions of complex differential equations. We are not really sure that it corresponds to any reality, so somehow one has to understand what the value of a simulation is.

Smale: I would say that the problem could be divided into two parts. One is: Is the model correct? In biology this is a superproblem, but in lots of other situations we know the model is correct. When we take equations from physics and try to study them on the computer, then the problem becomes: Is the round-off going to destroy the validity of what one sees on the numerical outputs? And that problem is there also for the biological equations. But in biology —or the places where one does not have a strong model— one has two problems. One is, as I said before, whether the model is correct; and the second is: Is the output reflecting the model? In a lot of the other applications, in engineering and physics, the model is substantially correct, while we still have to worry about the exponentiation of round-off error, and things like that, for dealing with which there is beginning to be some kind of science, but that becomes a substantial part of science itself. To what extent can one accept these outputs of a computer, even on models which are fine?

Wagensberg: Do you distinguish between simulation and calculation? I think you are confusing them.

Jones: I think that what Professor Smale is saying is that there is simulation where we do not really know the model, and there is calculation where we do. Probably there will be a fine line between these two. When we start to get to turbulent phenomena, there is something clear that the Navier–Stokes equation should be correct in that particular regime.

Smale: Navier–Stokes are not so bad.

Jones: Well, in a sense, that is one of the things that the computer calculations are testing.

Thom: I do not like computers. Essentially because of the fact that they replace any form by a set of pixels, so the discretization, digitalization, is giving up any kind of analogical thinking. I do not think this is progress. I think, on the contrary, it is a regression. But, of course, as I am alone in this kind of thinking, it will be difficult to justify my point of view.

Jones: It might be very difficult to solve my problem without computers. I just cannot do it, so I am rather happy to resort to computers, even if they are digital, to try and solve the problem which I could not solve otherwise.

Wagensberg: I will invite Professor Thom to justify himself a little bit more. Why do you think computers are a regression?

Thom: It seems to me that the process of digitalization could be justified inasmuch as we know that there is a sort of underlying continuous process, which has its own underlying evolutive mechanism of a differential nature, for instance. If this is not the case, if we just discretize for the pleasure of discretization, then I am not sure that this is really progress. Of course we can do it by sheer brute force. Then that is essentially the present-day philosophy of the use of this kind of modelization, like finite element methods and things of this kind. But I am not sure that this is really a successful progress for thought.

Jones: There is a beautiful example of this, which I learned when I was a student. Suppose we just wanted to solve something really simple, like Laplace's equation. If we have a nice circle, or maybe a square, there is a beautiful theory for how to solve this equation, and lovely formulas for the solutions given the boundary. Now suppose we just put a little dent in that square. Maybe we still want to solve it. However, all of our analytic apparatus falls apart. But sometimes one wants to know the answer. What is the displacement of the medium at a certain time, in a certain place? And this is just too hard to do by any analytic methods, yet the computer will give us a perfectly good answer. I do not see any reason to reject it on philosophical grounds. It is sort of a regression in thinking.

Smale: Yes, and against what Professor Thom was saying, let me give an example of where discretization has played a fundamental rôle in mathematics. We take the continuous object of a differentiable manifold, and we substitute for that the cobordism ring, which is a discrete algebraic object. It is a great insight. So there is a lot to say for discretization.

Thom: The relation of being the boundary of something is a continuous relation, so that we can generate a discretization by passing to the quotient modulo a continuous relation. There is no mystery in that. But if the continous equivalence class is arbitrary, like when we replace a continuous form by a system of pixels, I have doubts about the digitibility of such procedures.

Jones: I think there are results proving that continuous things can be approximated by discrete things. There are actually results saying that these programs that use discrete methods to estimate values, actually do converge in some sense, and we can get arbitrarily close in the perfectly discrete system.

Thom: In some sense, the belief in continuity, and in an underlying continuous process, is a deeply ingrained belief in our philosophy of science. It started with Galileo, essentially, or even earlier. But, mathematically speaking, as I explained in my talk three days ago, this was associated essentially to the fact that analytic continuation is our only sure way of extrapolation. To extrapolate a function in a kind of canonical way, there is only one way which is secure, namely analytic continuation. And analytic continuation is something which is infinitely differentiable, so we cannot avoid considering an underlying continuous process, if we want something which is justified. Of course, if we do not worry about justification, we can do anything we like, and success will be the only criterion.

Wagensberg: Professor Faltings is a specialist in Diophantine equations. Do you have an opinion on the continuity of nature?

Faltings: In some sense it is a matter of experience that we think nature is continuous, because it fits in with what we learn from youth; but, of course, we know from physics that things are not so simple, for example, that parity is violated, which we do not see on our level. In principle, it might be possible that nature is discrete, or something else. But, if we do mathematics, we make an abstract definition and have real numbers, and we can work with them, but I do not think it is really necessary, or clear, that the real numbers ultimately will describe nature.

Wagensberg: Do you think that everything can be described with a mathematical model?

Connes: I would like to come back to this discussion about real numbers, because I do believe that we receive a sort of intuitive idea about real numbers, and we do not realize it when we apply real numbers to physics. If we wanted to know a single real number, its digits would fill all the libraries of the Earth, and then we would only have a very partial knowledge about it. In physics, it is now clear that there is a certain length of the order of 10^{-33} centimeters, which is called the *Planck length* and beyond which, for very good theoretical reasons, it is impossible to decide what is a length and what is a point in space-time. So I want to stress that we receive from our youth a model —in fact, it seems that this model already exists in the brain at a very archaic level when we are born— which disappears around the age of one, one-year-and-a-half, and gets replaced by a better model, which is being acquired. But we have an archaic model about space and time, which allows very young babies to orient themselves, not to fall in holes, and so on. Now this model is, perhaps, one of the best conquests of man, in many ways, because we have a fairly good understanding of space and time, and this understanding improves. On the other hand, I want to stress that it is only a model, and when we think about coordinates of a point in space-time, the real numbers would require infinitely many decimals.

Let me explain this point, because I think it is interesting and everybody should know about it. There is the uncertainty principle, which tells us that if we try to pin down a point, try to know its position very precisely, then it has a tendency to escape, and the speed with which it escapes is inversely proportional to its mass. Thus, if we want to pin down a point, we should try to pin down a point which is extremely heavy. Now if we put all the mass of the Earth, and concentrate it in a small region, then we find out that, in a radius of about 3 centimeters around this region, because of the gravity, the photons, the light rays, cannot escape. So this is another effect which prevents us from increasing this mass to a sufficiently large size, so that we would know the position. If we play these two facts against each other, we find that there is a length, which is of the order of 10^{-33} centimeters, beyond which the very notion of a point in space-time is meaningless.

So I believe that what we have achieved so far is to have a model, this model with real numbers, which is striking by its property of being self-similar. This model is a good approximation of what the universe is, or what we understand it to be, but it is an approximation which is very much like a tangent to a curve, because we impose on it this notion of self-similarity. And I believe that when we reach higher and higher energies (or similarly, smaller and smaller scales) this model will evolve, and at some point it is fairly clear that it will depart from the usual notion of real numbers. It will depart in the sense that it will give a picture which is no longer self-similar, and which incorporates this better knowledge that we shall have acquired.

So, I really want to emphasize even the simple notion of real numbers, even the way they appear in physics, because in quantum mechanics they appear in two different ways: they appear also as probabilities. There is a very striking fact in quantum mechanics, namely that a probability is a real number, because the proportion of events which occur can be counted. If more and more experiments are done, a better and better precision is obtained. So these probabilities occur as the square of the modulus of complex numbers, and these complex numbers are amplitudes, coefficients of vectors in a vector space, and they have a fairly different nature. So I think that, even this very simple question of how real numbers occur in space-time and in physics in general, is a fairly difficult question to discuss, and the question which shows us that we are bound with models, and nothing more.

Wagensberg: Anyway, your answer suggests to me a definition of nature: for example, the limit of the models when the time of research tends to infinity.

I will now ask Professor Smale a concrete question about his own field, and about new fashions in science and in mathematics. For example, chaos is now a new interesting idea. We know that this field began, a long time ago, with the mathematicians. Then it entered physics (some preliminary expressions like the dissipative structures in thermodynamics). And now it is a big field that is invading physics, chemistry, biology. What is your opinion about that?

Smale: Well, first of all, I take issue with many of my colleagues in dynamical systems. I think that chaos is very important for science to get at. After quantum mechanics and relativity, it is a new challenge to the determinism of Newton and Laplace. It shows the weakness of models which people believed in the last century, where the future was determined by initial conditions. It shows how this breaks down on a new level, not on the macro-scale, nor on the micro-scale, but just on the everyday-scale, where determinism fails. And that lack of determinism lies in the equations themselves. So I think that there is a kind of revolutionary aspect of chaos. And it illustrates to some extent another difference in mathematics, an advantage that mathematicians have over physicists. I think the first way that chaos was found was more in mathematics than in physics. For example, there is Poincaré's finding of homoclinic points. And that came about partly because mathematicians were not so much tied to the particular equations. They had a bigger universe from which to study mathematics. Even in the last decades, one sees physicists, and many mathematicians as well, shying away from the general differential equations; they say "General differential equations are irrelevant; one needs to study the particular equations of physics, or engineering, and these are not generic." On the other hand, mathematicians have a very wide universe; they can look at all equations very easily, and in that framework, I think, these ideas of chaotic dynamics were really developed, mostly by mathematicians, before physicists. And I think this illustrates something that mathematicians can lend to science. It is some kind of character that mathematics has, namely the idea to look at some wide universe of mathematical models.

Wagensberg: It is interesting to realize that we are living in times where popular science is very closely following what is happening in science. For example, we are planning an exhibition on chaos in this museum.

Professor Thom, what is your opinion about chaos, this new fashion in physics and in many fields of science?

Thom: My point of view about modern physics is the following. In Aristotelian physics, one believed one was looking at natural phenomena, allowing for accidents, and science could consist only of natural phenomena. Accidents should probably not be the object of science. Then Galileo said that when we throw a stone then it has forced motion, followed by the natural motion of falling down to the earth. The forced motion was somewhat an accident for Aristotle. For Galileo, however, both motions had the same equations, and so he decided that it was the same motion. Now, science has always tried to get rid of accidents, recover accidents, and subsume accidents inside more general and more complicated laws. And this is the process which is still continuing in fundamental physics. Now people are quite happy to consider phenomena involving resonances about 10^{-32}, and if we follow this way of thinking, that we can extrapolate this process of recovering accidents, we would arrive at the conclusion that the stability of the world is founded on extremely unstable phenomena. This is extremely difficult to believe, at least for a philosophically-minded man. And perhaps you, as a physicist, are quite happy to recover accidents, but it seems to me that it is not really very acceptable to believe that one can indefinitely continue towards the elucidation of the infinitesimally small, and the infinitesimally short from the point of view of time. There is no limit which could be considered in that respect.

Wagensberg: I propose now to provoke all of you. You were comparing mathematics with other sciences. I will make a comparison between mathematics and art. I know that many mathematicians are happy with that, and others are not. So I will make a brief list of similarities, and those of you who are unhappy can list differences. In physics you need to be objective, you need intelligibility, and you need a kind of matching with experience. On the other hand, you have mathematics and the mathematical reality, and you have also art. We can state that, for example, mathematics, like art, is more a personal creation than physics. This you can see very easily if you see the number of people who sign a paper in physics. You can see 10, 15 even 30 or 40 people signing one single paper in physics. In mathematics, I think the mean value is one person signing one paper. Another point is that the work is more attached to its creator in the case of mathematics or theoretical physics than in the case of biology, chemistry or physics.

Aesthetics is also very important. Even in theoretical physics; for example, Dirac said that elegance can decide which of two different equations, with the same length and the same merit, explain some real situation. And aesthetics is also important in art.

So would you like to be compared with, for example, a poet?

Faltings: I think you make the wrong analogy. What you really distinguish are things which require a high degree of organization, like organizing a particle accelerator, and making experiments there, and things like mathematics, which you can do alone. For example, I would consider the cathedral in this town a work of art, and it was certainly not built by one person, but by many persons. Also, it is partly because people in art do not give credit to their helpers. For example, when Christo packages a whole palace or something, he does not do it alone, there are certainly other people helping him. So in this respect I would not think that there is a big similarity between mathematics and art.

Wagensberg: In both, the creation does not depend on the people who are helping.

Faltings: Yes, but in physics there is also one leader and people he gives tasks to.

Wagensberg: I protest. Not in the same way as Christo with his helpers.

Faltings: This I cannot know. Also, physics is not only particle physics, but there are also other physicists, and they work more alone, so I do not think that there is such a distinction.

Thom: I would like to say a few words about aesthetics in mathematics, without speaking about art itself, which I think is a somewhat different question. Many people claim that there is a lot of beauty inside mathematics itself, but that is not my own experience. Nevertheless, I believe that there is a sort of general structure describing how beauty occurs in mathematics. For that, one has to classify mathematical objects in a plane spanned by two axes. One axis, a vertical axis going down, would be an axis describing how a structure is constrained. We start with phase structure, let us say at $z = 0$, and then go down. The more we go down, then the more we get constrained structure. The other axis, the horizontal axis, goes from discrete to continuous, starting with the natural numbers $\mathbf{N}$. Then, going down along more or less the main diagonal, we get $\mathbf{Z}$, $\mathbf{Q}$, all the fundamental objects in mathematics —functions, differential calculus, etc— and one has the feeling that there is a sort of central general scheme of mathematics going diagonally inside this plane. Now, beautiful objects occur below this main axis. These are objects which are "more constrained than necessary" and exhibit a lot of internal symmetries. So they are gems, so to speak. We get fewer gems below the fundamental branch of mathematics; and above we get the disordered objects, broken structures, and objects which are rather ugly. A typical example is the following: We start with algebraic objects, defined by algebraic equations (let us say, with real coefficients), and then we want to project down a real algebraic set. Then we only get as an image, in general, a semi-algebraic set, so we have to introduce inequalities. Each time we impose some sort of relatively simple constraints on such a beautiful object, then we get some sort of ugly object. It is an *objet moche*, as I say in French. That is, we get semi-algebraic sets, and then semi-analytic sets, sub-analytic sets, and so on. So above this branch we find a lot of ugly objects, and there are people who like to work with these objects. So mathematicians, roughly speaking, can be divided into two classes: those which work on the beautiful side and like beautiful objects, and those who like to go into the mud of ugly objects. This is perhaps a kind of generic description of mathematicians.

Jones: We had a very interesting discussion last night about the question of progress in arts and in mathematics. I do not want to go through it again, so let me resume. No one will deny that there is progress in mathematics, but I contended that in some sense there was no progress in art. My example was: Take a piece of Bach; one can listen to that today, and it can be completely fulfilling. Whereas we argued about Gauss (we should have perhaps argued about someone a little earlier). Basically, concerning the old theorems we look back on, the proof was very nice, very cunning, but we know how to do it better now, we would not actually do it that way. So the question which I want to raise is: It is all very well for us to say this now, but we are not professional musicians; maybe a professional musician is so damn sick of hearing Bach that it is sort of the same thing. Do musicians get really mad at all this stupid general public for wanting to listen to all this Bach again, when they should be listening to something that was written in the twentieth century? I think that is an intriguing question.

Connes: I just want to make some comments. There is probably some artistic component in mathematics, but I think it is quite important, though, to try to pinpoint in what sense. The mathematical reality is similar to what physicists or biologists study. I once had a long discussion about the nature of mathematics in comparison with biology, and I believe the following. When Watson and Crick discovered the structure of DNA, nobody would have told them that they invented it, and that it first existed as a configuraton of their neurons.

I think one has to be quite specific, and say that there are two different attitudes that a mathematician has towards mathematics. And one of them is, of course, he has to create tools, exactly as Watson and Crick had to have the electronic microscope, in order to observe things. So a mathematician has to create tools, and he does a work of creation at this point. On the other hand, until a mathematical tool has been used to unveil a part of this mathematical reality that I am talking about, it is not really accepted by the mathematical community. It is considered, but it is not really accepted until it has been successfully used. This is something which is called a "breakthrough," and we understand what that is. So I think it is a mistake to talk only about creation. I think there are the two aspects, and these two aspects are fairly clearly separate when talking about physics, or when talking about biology. However, in mathematics, they are more mixed together, and are more difficult to separate, yet they are present. One can of course have some comparisons of mathematics with art from the outside. Nevertheless, one cannot reduce it to an art; it is quite different from that standpoint.

Wagensberg: Of course it is different; I said only that it is closer than physics.

Smale: I would like to take issue with what Professor Jones said in his analysis of the importance of something that Gauss did. For me it is not so much the proofs in mathematics that are important. Certainly the theorems of Gauss have much shorter proofs today. But what is valuable, and what is beautiful in Gauss —what has persisted today— are some of the ideas of Gauss., Some of the ideas of Gauss stand today as beautiful as ever, and are not obsolete. Of course, proofs are much more efficient now, but to me, the proofs in mathematics are not the primary things. It is the ideas. Proofs are there, but the beautiful and great things are the ideas, and many of those ideas of Gauss persist today.

Wagensberg: Just like Bach, you mean.

Connes: I want to refer to what Professor Thom said about the division of mathematics into two parts. Everyone would be very tempted to say that they are on the aesthetical part, and not on the other part. On the other hand, I think it is very interesting as a question, because, for instance, if we think about the origin of Riemannian geometry, at the beginning there was Euclidean and Lobachevski geometry, and these, of course, can be qualified as being gems, if considered as a part of general Riemannian geometry. When beginning Riemannian geometry, it is tempting to limit ourselves to spaces in which rigid motion is possible (symmetric spaces, for instance) which one could qualify as being gems. On the other hand, when we look at the development of general relativity, it is absolutely necessary, in order to formulate the theory, to have at one's disposal arbitrary Riemannian metrics. So, one has to be very careful in qualifying the various subjects, precisely because of examples like this.

Wagensberg: Before we give the audience the opportunity to intervene, I would raise a last question that I think could be very interesting for us here in the mathematical community in Spain. What can be done to promote mathematics in a special community?

For example, it is very clear that it is also a question of tradition. Yesterday, Professor Smale said that the case of Brazil is very important. And when this tradition is broken, then it is very difficult to recover again, at the same level. This would perhaps be the first part. The second part is: What do you think about the quality of research in mathematics? We live in special times, where researchers in the university are under stress because they are measured by the number of papers, by the weight of their production, and perhaps we do not care so much about more fundamental problems that entail more risk. So the twofold question would be this: What can be done to promote mathematics in a particular community? And what is your opinion about the quality of the research, whether there is some inflation in mathematical production?

Smale: I have been to Brazil eight times now and I find it a good example where mathematics came from very small seeds thirty years ago, and now it is among the Third World's most healthy mathematics. Very exciting mathematics, especially in Rio de Janeiro. I do not know exactly what the mechanisms are of all that. Certainly the main things have to do with the energy of Brazilian mathematicians, but I think the support they have received has been important also. I do not think it is so good to try to intervene and do things in other countries, but when one sees the seeds developing, it seems to me very useful and important to give those kinds of things support. So, I see the rôle of mathematicians like myself as being more supportive in developing mathematics in other parts of the world.

Faltings: There is a question of inflation in mathematics, and that people have to write too many papers, and that people do not do the fundamentals. I do not see this as such a danger. There is always much more bad science than good science. To get one good scientist, you have to support nine bad ones, because you cannot know a priori which of those ten will be the good one. I think this is natural. And also, there should be some measure of productivity for a mathematician. We cannot expect the general public to let us do what we want to do, without caring of whether we work or not. And of course people count papers; but they usually count them with a weighted count, depending on the journal where they appear; one should not just count papers, but if one also looked at the quality, then I think this is not so bad. And as to the quality of mathematics in general, I think we are no worse than in ages before us.

Wagensberg: The debate is now open to the floor. Any question will be welcome to the round table. Some questions or comments?

Brown[1]: I have a question. In some subjects, like sociology, for example, or economics, there is much discussion about the methodology of research. And there is much consideration given in the training, say, of students, even undergraduates, to this question of the methodology of research, and how it progresses. Do you believe that there is some kind of methodology of mathematical research which can be communicated, can be discussed, and which would be of value to students?

Jones: I think that, to a large extent, that is what conferences are about. I believe it is very important for students to go to lots of conferences, and be exposed to all kinds of

---

[1]Full names and affiliations of participants from the floor can be found in the List of Participants.

different mathematicians. Of course, this is at a graduate level. I think this is the basic communication ground for research methodology. I would hate to try to axiomatize, or tie down, the methodology, but I think it gets communicated primarily at conferences.

Smale: To me, the most important methodology is the methodology of doing mathematics, of solving problems, and, to me, the importance of mathematical lectures is to create an atmosphere where students try to solve problems. This may be the main importance of mathematical lectures, i.e., to set this environment where people are actually doing mathematics, by trying to sort out things and solving problems in a very broad sense. Perhaps organizing mathematics on paper. That is the methodology, and I think the lecture-style of teaching is not such a happy situation, because too much time is spent listening to mathematics, which is too passive. Doing mathematics seems to me the primary thing, but one has to have the atmosphere which comes from the lectures.

Dou: Earlier the question arose whether the world is discrete or not, or at least it seems to me that Professor Thom questioned the possibility of discretization. You said, if I understood correctly, that it is substantially distinct from reality to take for valid, say, finite equations, and not to consider the real problems which normally are not possible to transfer into an analytic continuation. But I would rather believe that the world, the real world, is not continuous. The real world is discrete. You imagine space as a continuous manifold, but really, at the deepest level, at some point, we have discrete items only. For instance, quantum is discrete, length is discrete, mass is discrete. I have the same reasons for saying that the world is discrete as Xenon of Ancient Greece.

Thom: First of all, if you admit that the world is fundamentally discrete, you cannot deny that your own subjectivity is continuous. Or do you deny that? Do you deny that you feel the flux of time as a continuous flow?

Dou: I feel that the flux of time is continuous. But I feel also that moving pictures are continuous, and we know they are not. There are other methods of accounting for continuity, but nothing in itself, not a single real world entity is continuous. What is continuous is the take-over of the action. The picture in the motion pictures is continuous.

Thom: Even if our feeling of continuity is an illusion, then you have to accept that being an illusion is already a kind of existence. I agree that the classical example is when you see a film in a cinematograph: on the screen you have the feeling of continuity, but you know that on the film you have a sequence of discrete pictures. And your opinion would be that it is the same in general for our feeling of time. But then I would ask you, how is it that we have the illusion of continuity? That is the problem. How is it, if there is not an underlying continuum which contains the continuum? I wrote in a meeting last year precisely about the question of continuum in mathematics, and I think there is a very simple argument in favour of the continuum, with respect to the discrete. Namely, consider two concepts, let us say A and B, and say that A is *ontologically prior* to B if any entity which has quality A, has automatically quality B. Then if we compare that to a continuum and discrete, we will rapidly be aware of the fact that there are continuous objects which have discrete qualities, while we cannot have a discrete object which has continuous qualities.

Dou: Wait a moment. I cannot say by any means that everything in the universe is discrete. I say in physics everything is discrete, but not in the universe. The universe is in some sense continuous. There are plenty of unknown things that in some sense create, not only the illusion of continuity, but some kind of continuity. However, I cannot imagine the physical objects as being continuous.

Thom: When you refer to a physical object, you mean an object which can be scientifically described. I would accept that, in any kind of description, we have a discrete element, because a true continuum has no points. We are unable to specify anything in the continuum. The continuum is something which cannot be described. It is a sort of unsayable. It is a world in which one lives outside of symbolic description. But nevertheless, it exists, despite the fact that we cannot describe it in any sense.

Jones: I think one of the great things about mathematics is that we have sort of solved this problem. We invented the real numbers, and gave axioms, and defined them, and so on. So we do not have to get too upset about discrete and continuous nature.

Wagensberg: But all numbers in nature, in physics, are rational numbers.

Pfenniger: We actually have a kind of model for bridging the gap. Simplicial complexes have the power to model quite a bit of the continuum, but nevertheless live very much in the discrete world.

Smale: Maybe a comment on the discrete versus real. I think one resolution in physics and mathematics is the idea that there can be many idealizations of physics. There is no unique idealization. There is relativity. There is Newtonian mechanics. There is quantum mechanics. Each of them describes reality, but they are different idealizations. There can be discrete idealizations. There can be continuous idealizations. And there is no unique idealization. So I think somehow this may be not a real issue. It is not an important issue, one versus the other. There are many idealizations. We need idealizations, and we get insight from various idealizations. To me that is a resolution.

Connes: I just want to make one comment: I ask myself whether there is an interesting aspect behind this question, because one can certainly get away by saying that what is important are models. We cannot say that the universe is discrete, or that it is not. What we can decide is whether our models are successful or not. And it would be wrong to say that in quantum mechanics everything is discrete. This is absolutely wrong. There is indeed some element of discreteness in quantum mechanics; but, on the other hand, the real issue comes when we try to understand the duality between the corpuscular aspect. For instance, Newton began by thinking that light was made of corpuscles, just because there were shadows. And then people, by doing a lot of experiments, understood that there was a wave-like aspect in light, which would be very difficult to describe as being discrete. And then, gradually, it was realized that the two aspects were present, and that, when taking very long wavelengths, (like radio waves, for instance) they go around obstacles, so that they cannot be understood as photons or particles. So I believe that the two aspects are present. They are present in the models, and we should not qualify the universe. What we have is a fairly good model, which works because of the differential calculus of Newton, which allows us to compute with the continuum. What is important is to be able to do something with it. Now one way to show that these two aspects are not different from each other is to say the following. Often we have a perception which is only two-dimensional for phenomena in three-dimensions. For instance, an ice-cream cone: If you look at it from the side, you see it as a triangle, and if you look at it from the top, you see it as a circle, as a disk. What happens is that reality is so complex that, in some cases we see it as made of corpuscles, we see it as discrete, and in some cases we see it as a continuum. And these two aspects cannot be disentangled from each other. I think it is very important not to qualify them as being elements of nature. They are just elements of a model.

Faltings: I think we should not worry too much about reality, because we make mathematical models precisely because we do not want to test everything in reality, but only against our axioms. But I think another interesting aspect is that, in mathematics, there was a similar crisis about discrete and continuous, namely when Hermann Weyl asked "What is a real number?" Usually we say "You just write down the decimal digits, and then you go on at random." And then he would respond "This is not allowed. You should be more careful." And so, to say what the square root of two is, we cannot say "You take the squaring function: it is 1 at 1, and it is 4 at 2, so in between there must be a value where it is equal to 2." We have to give a procedure describing how to get those digits, which we can do. And so, I think, in some sense, we are coming back to computers. In the old times, people would do more concrete things. They would not talk about an arbitrary manifold. They would talk about this manifold, or that manifold, and study it. And somehow, this aspect of taking abstract surfaces, which is basically due to Riemann, is somehow recent. I know there are people who did not like it, like Siegel, who always complained that the computational aspect of things, and the small concrete things, got lost, starting with Riemann, going on with Hilbert, and so on. So it may be that we are engaging in an old debate here.

Wagensberg: More questions?

Burenkov: One of the titles of our conference is "The Prospects of Mathematics," and it would be very interesting to know the opinion about our standing in mathematics. What are, in your point of view, the general prospects of mathematics with respect to the two following aspects. The rôle of mathematics in science and society in general: Will it grow or not? And what directions in mathematics itself will have the highest growth, in the near or distant future?

Thom: Well, I think there is absolutely no way of making any kind of prediction about the way mathematical research will develop. All considerations of this kind have no object.

Castellet: I would like to come back more to mathematics, and especially to conjectures. Many of you have been involved, or are involved, in some of the famous important conjectures in mathematics —the Fermat conjecture, the Poincaré conjecture. I would like to know what is your opinion on the development of mathematics through trying to solve such conjectures. And, what will happen when one of these famous conjectures is solved? Probably nothing? And another question about conjectures: There are some famous conjectures, as I mentioned, but now in recent years a lot of conjectures have appeared. Everyone has his own conjecture. And three months later someone finds a counterexample to this conjecture, or proves the conjecture. Well, probably the word "conjecture" is no longer correctly used in many papers.

Faltings: Usually the experience is that if a big conjecture is solved, it is usually the end of something, and it is a sort of let-down. Things were much more interesting before it. Part of the progress of science is proving those conjectures, but this only gives a partial view, because what is at least as important is doing a theory which is usually developed in solving these, a theory which allows to make new conjectures and new questions. And so I think that, if you put too much emphasis on conjectures, it somehow limits this view of other important parts of mathematics.

Noy: I do not want to ask about the prospects, but about what you think is the current state of mathematics. For instance, if you read the last issue of the Notices of the AMS, you see a conference on number theory and dynamical systems, which is something which could be surprising. Is this an indication of a certain kind of unification? That is, is there more relation between the different branches of mathematics? Do you think this is the tendency?

Smale: I think that both things are always going on at the same time in mathematics. There are diversifying forces, some of which are good, and there are unifying forces, some of which are good. I think this is happening very much nowadays, because more mathematics is being done nowadays. But I believe that both of these tendencies are found, as well as forces giving sort of a whole life of mathematics.

Wagensberg: Professor Connes reminds me that a very interesting topic is the new fashions and the marketing that is done to popularise mathematics. For example, even in the interior of the mathematical community, it is a good thing to write popularising books. We have, in the case of chaos, the example of Glick. I think that a lot of scientists became aware of chaos because of a book written by a journalist. This is a very interesting case. What do you think about that? We coin new terms like "catastrophe," like "fractals," like "chaos." Is this a positive thing?

Connes: Should I answer a question which I raised? What I had in mind was the question of unification of mathematics, and the trend of string theory, and the way a lot of people who were doing theoretical physics became like pure mathematicians. This was qualified as being one very strong interaction, not only inside mathematics, but between physics and mathematics. And I believe that one should think about it more closely, and try to qualify it more accurately, because what it seems to mean is that a number of theoretical physicists migrated from theoretical physics to mathematics, and not much more than that.

I am making a provocative statement, but I believe that it does depend very much on the way things are presented and on the number of people who work on the topic, because there were, about fifteen years ago, extremely strong relations between quantum statistical mechanics and pure mathematics, which were not so well advertised, and which did not attract crowds of that magnitude. And this depends very much on this type of realization.

Wagensberg: What about the level of volume that a certain field reaches?

Connes: There are fields where it is possible to access very directly, and where it is possible to write papers and to publish. But there are fields like, for instance —I would take an example which is really completely the opposite— constructive quantum field theory, which are extremely hard fields, in which if one wants to do anything, one has to first work for four or five years without producing anything, just understanding the basic techniques. In the States, for instance, it is very difficult to attract people in such a field. Why? Because it is not giving a big payoff from the start. Nevertheless, I do not believe that it is less important, not at all.

Brown: I think one of the questions which concerns all of us is the status of mathematics and its, in some ways, lack of appreciation by public and government. For example, many people believe that supercomputers are important, but do not understand that mathematics is at the basis of this. Or, for example, important practical applications, like X-ray tomography, are not generally realized as having an important and necessary mathematical basis. Do you feel that mathematics is properly appreciated? Is it not properly appreciated? And, if it is not properly appreciated, do you think that the senior mathematicians speak up enough for mathematics, as a subject?

Smale: I think the answer is not so much what the lobbyists are doing. The American Mathematical Society is doing more and more lobbying —congresses for funds, and things like that. So there is this lobbying aspect which I will not condemn, but on the other hand, to me, a more important direction would be for mathematicians to learn to speak a more general language and get out of the language of their narrow subject, be able to speak to wider audiences within mathematics, and even wider audiences in the scientific community, by putting a lot of effort into expressing their mathematics, even their research mathematics, in terms with which we can reach a much broader audience. It seems to me that this is the main constructive way of dealing with that situation —the idea of putting a lot of effort into trying to communicate mathematics to a much wider audience. That seems to be much more important than the idea of lobbying for more money for mathematics.

Aguadé: Well, since our speakers seem to prefer philosophical questions to concrete ones, let me try to ask them about how it is possible to measure the significance of a result in mathematics. For instance, there is probably a Fields Medal waiting for the one who solves the Poincaré conjecture, but there was no Fields Medal for the four colour conjecture, for instance. So what makes the solving of the Poincaré conjecture more important than some other problem? How do you measure it? If mathematics is not like art, there should be a criterion to decide what is important, and what is not so important. For instance, imagine that the solving of the four colour problem, instead of using a computer, had used the whole machinery of algebraic topology, and quantum groups, and whatever you can imagine. Would that make the proof more interesting? Would it then deserve a Fields Medal? So how do you measure what is important and what is relevant in mathematics, and what is not?

Jones: I believe the answer is quite easy: yes. If the four colour problem had had a nice proof, then it would have been much more significant. The only true measure is what happens with time. Things get sorted out eventually. Necessarily, in any awards, there are going to be mistakes, and more significant things will get ignored. That is just inevitable.

Smale: I would be glad if you could answer the question "how to measure significance in mathematics." It does not have an easy answer.

Jones: That was why I answered the other question.

Faltings: Temporarily, significance is just measured by popularity, and, in this case, the four colour problem was not so popular. For example, the Poincaré conjecture in dimension four, even in higher dimensions, got a Fields Medals, partly for the theory which was developed and could be used for other things, and partly because it was more in the mainstream of mathematics.

Thom: If you put the question of significance on a general level, for almost any science, then it would be extremely difficult to evaluate the significance of a result. In biology, for instance, it is practically impossible to evaluate the significance of a result unless it is associated to a really specific practical, or pragmatic, purpose where you can have success more or less immediately. There is practically no sheer theoretic level in biology. From the point of view of mathematics, I should think that the general consensus among mathematicians is relatively strong. There might be, of course, divergences of opinions, from one specialist to another, but, roughly speaking, I suspect that there is a fairly objective consensus.

Grundbaum: I wanted to make a sort of political statement, a sort of corollary of something that Professor Smale was saying earlier. If mathematicians are going to be able to go out and talk to the general public, or talk to other scientists, or to congressmen, or even to other mathematicians, I think there is an implicit need to make sure that people get a rather broad education, which conflicts with the way that we normally produce a young research mathematician. A person has to go very quickly into a field, and does not really have much time to learn about anything. This is really going to make an impact on what sort of statement a mathematician can go and make in front of people that are willing to contribute more effort or more money to mathematics. For instance, if we spend most of our energy debating about reality as defined for a discrete or a continuous model, I think congressmen are going to look the other way, and they are going to be pretty bored with the whole question. They are not really interested in that kind of thing at all. The whole point of mathematics is that it can bridge among all these different things, as some other people have said. If I have to end with a question, the question is: Should we do anything at the level of, say, college education —or whatever you call it in other parts of the world— to make sure that the formation of young mathematicians looks more like it used to be fifty years ago, when people were forced to learn some of the other sciences? Or should it be more the way that it is becoming now, where you can become pretty self-focused if you want? In general, is there anything we should do, anything that concerns us here, about the way we treat, say, undergraduates, in the States at least? Maybe other parts of the world do not have the same problems. Yet I suspect they do.

Faltings: I do not think we can force people to learn what they do not want to learn. So this, I think, should take care of undergraduates in the States.

Grundbaum: Is that really a defensible position, that we cannot force people to learn what they do not want to learn?

Faltings: I do not think that is a position. It is a fact.

Grundbaum: Is it a dangerous fact, in terms of mathematics?

Bokor: We do force people to learn. We set a syllabus; we do not teach Gauss; we do not teach the history of mathematics. And if you are a kid going into research, you are told to read the latest articles first, and very rarely do people actually go and read the originals, the masters. That is seen to be a waste of time because there is no payoff. So we do force people to learn things they would not of themselves try to learn. Just to stay alive as mathematicians, you have to do things.

Faltings: I should say I have never read Gauss myself. I think there is progress in mathematics, and you do not have to read Gauss any more, because what he has done has been subsumed in newer treatises, and maybe we can force people to learn something, but they certainly will not like it. Usually people want to learn what they can use, and the average college student can exist without being able to add two-thirds and one-fifth. We may teach it to him, but he will soon forget it. On the other hand, we cannot expect the general public to learn our trade. For example, there may be Finnish linguistics, or some other area, where people also think that the general public does not know about them. But there is a certain limit to what people want to know.

Venakides: I think the essence of the last question was that we all agree that mathematics is beautiful and useful; but, how should we make undergraduates perceive this fact?

Faltings: Only by our personal examples, I think.

Brown: Earlier Professor Connes referred to Watson and Crick, and in the television program on Watson and Crick, the young Watson said "Big questions get big answers; I want to know: What is life and how does it replicate itself?" Do you feel there can be, or there are, questions of a comparable type in mathematics?

Faltings: I give you an example: "What is motives?"

Connes: There is one thing which I want to say, however, about the discussion that we had on evaluation in mathematics, and I would really stress that, of course, there is problem-solving which is quite important; there is also elaborating theories which generate problems; but I believe there is something else which is quite important and which has nothing to do with problem-solving or problem-generating. It is to show that things are far better than what one would believe from the outset. This is really what makes a distinction between solving the four colour problem, or solving the packing problem, which people say can eventually be reduced to computer calculation. Once we have the answer, then ... what? Compare this with exploring some part of the mathematical reality, which, from a distance, would have appeared to be strange, without any structure, ugly if you want, and unveiling in fact a lot of structure, a lot of richness, in this part of mathematics. I think this is really what makes a distinction. It is not even a question of solving or not. I think this really makes a distinction at the evaluation level. That is the opinion I have about this.

Now, if you ask about fundamental questions, of course there is this question of motive, which is not only true in algebraic geometry. It is also true in the subject where I work. In the classification of Bourbaki, algebra, analysis, and all that, were put on very different footings. However, in the last twenty years, some theories have appeared, like K-theory, algebraic K-theory, which completely unify strata which normally would have been considered as being totally disjoint. I was struck, for instance, by things in the talk of Faltings. He explained that one way of looking at the geometric proof was to formulate the right algebraic K-theory, formulate an index theorem with the Chern classes, and so forth. This is exactly parallel to the type of development which I was trying to explain in noncommutative geometry. So what I would try to say is that there might be such transverse theories which, in fact, make totally obsolete any attempt that Bourbaki would have made to set the future problems of mathematics or anything like that. And I think we have the experience that the developments are always completely surprising with respect to whatever one could have tried to foresee.

Wagensberg: Thank you very much for your opinions, and for your short and long answers.

Transcribed from the recording of the session by Warren Dicks.

List of Participants

LORENZO ABELLANAS
Universidad Complutense de Madrid
Spain

PATRICK AHERN
University of Wisconsin
USA

ANTONIO ALCARAZ
Universidad Politécnica de Sevilla
Spain

LLUÍS ALSEDÀ
Universitat Autònoma de Barcelona
Spain

MONTSERRAT ALSINA
Universitat de Barcelona
Spain

CONCHITA ARENAS
Universitat de Barcelona
Spain

PILAR BAYER
Universitat de Barcelona
Spain

MANUEL BENDALA
Universidad de Sevilla
Spain

BEATRICE BLEILE
Zürich
Switzerland

IMRE BOKOR
Zürich
Switzerland

JIM BRENNAN
University of Kentucky
USA

RONNIE BROWN
University of Wales
United Kingdom

VICTOR BURENKOV
Friendship of Nations University, Moscow
Russia

URS BURRI
Muttenz
Switzerland

ÀNGEL CALSINA
Universitat Autònoma de Barcelona
Spain

JAUME AGUADÉ
Universitat Autònoma de Barcelona
Spain

AURELI ALABERT
Universitat Politècnica de Catalunya
Spain

RICARDO ALFARO
University of Michigan
USA

CLAUDI ALSINA
Universitat Politècnica de Catalunya
Spain

BORIS APANASOV
Siberian Branch Academy of Sciences
Russia

FEDERICO BARTOLOZZI
Università degli Studi di Palermo
Italy

ROBERT BÉDARD
Université du Québec à Montréal
Canada

WOLFGANG BISCHOFF
Universität Freiburg
Germany

AGNIESZKA BOJANOWSKA
Uniwersytet Warszawski
Poland

LUIS J. BOYA
Universidad de Zaragoza
Spain

CARLES BROTO
Universitat Autònoma de Barcelona
Spain

JOAQUIM BRUNA
Universitat Autònoma de Barcelona
Spain

JOSEP M. BURGUÈS
Universitat Autònoma de Barcelona
Spain

CLAUDI BUSQUÉ
Universitat Autònoma de Barcelona
Spain

ROSA CAMPS
Universitat Autònoma de Barcelona
Spain

M. ANGUSTIAS CAÑADAS
Universidad de Granada
Spain

CARLES CASACUBERTA
Universitat de Barcelona
Spain

MANUEL CASTELLET
Universitat Autònoma de Barcelona
Spain

SALVADOR COMALADA
Universitat Autònoma de Barcelona
Spain

LUCÍA CONTRERAS
Universidad Autónoma de Madrid
Spain

MANUEL CUYÀS
Olimpíada Cultural, Barcelona
Spain

ROSARIO DELGADO
Universitat de Barcelona
Spain

ALBERT DOU
Universitat Autònoma de Barcelona
Spain

ALBERTO FACCHINI
Università degli Studi di Palermo
Italy

MARIA FERNANDA FARINA
Università degli Studi di Palermo
Italy

ROSA FERNÁNDEZ
Universidade da Coruña
Spain

MARIA JOÃO FERREIRA
Universidade de Lisboa
Portugal

MARIO ALFREDO FIORAVANTI
Universidad de Cantabria
Spain

GERHARD FREY
Universität-GHS Essen
Germany

ARMENGOL GASULL
Universitat Autònoma de Barcelona
Spain

JOAN GIRBAU
Universitat Autònoma de Barcelona
Spain

FERNANDO GRASA
Olimpíada Cultural, Barcelona
Spain

ALBERTO GRUNBAUM
University of California, Berkeley
USA

ANTONI GUILLAMON
Universitat Autònoma de Barcelona
Spain

DONALD HARTIG
California Polytechnical State University
USA

JOAN JOSEP CARMONA
Universitat Autònoma de Barcelona
Spain

CARME CASCANTE
Universitat de Barcelona
Spain

VICENTE CERVERA
Col·legi Universitari de Castelló
Spain

ALAIN CONNES
IHES
France

JULIÀ CUFÍ
Universitat Autònoma de Barcelona
Spain

M. CHAMARIE
Université de Lyon I
France

WARREN DICKS
Universitat Autònoma de Barcelona
Spain

RAMON ESPELT
Olimpíada Cultural, Barcelona
Spain

GERD FALTINGS
Princeton University
USA

MERCÈ FARRÉ
Universitat Autònoma de Barcelona
Spain

JOSEP FERRER
Universitat Politècnica de Catalunya
Spain

THOMAS FILK
Universität Freiburg
Germany

ERNEST FONTICH
Universitat de Barcelona
Spain

JAUME GARCIA ROIG
Universitat Politècnica de Catalunya
Spain

OLGA GIL
Universitat de València
Spain

MAREK GOLASINSKI
Uniwersytet Mikołaja Kopernika, Toruń
Poland

INDORINA GRAZIA
Università degli Studi di Palermo
Italy

JORDI GUÀRDIA
Universitat de Barcelona
Spain

FERNANDO GUZMÁN
SUNY at Binghamton, New York
USA

NÁCERE HAYEK
Universidad de La Laguna
Spain

FRIEDRICH HEGENBARTH
Università degli Studi di Roma
Italy

FERRAN HURTADO
Barcelona
Spain

STEFAN JACKOWSKI
Uniwersytet Warszawski
Poland

MARIA JOLIS
Universitat Autònoma de Barcelona
Spain

MARIUSZ LEMANCZYK
Uniwersytet Mikołaja Kopernika, Toruń
Poland

MIGUEL LOBO
Universidad de Santander
Spain

JORGE LÓPEZ
Universidad Politécnica de Madrid
Spain

ANA LLUCH
Col·legi Universitari de Castelló
Spain

MARTA MACHO
Euskal Herriko Unibertsitatea
Spain

TERESA MARÍ
Universitat de Barcelona
Spain

CELIA MARTÍNEZ
Universidad Complutense de Madrid
Spain

PAOLA MISSO
Università degli Studi di Palermo
Italy

RAIMO NÄKKI
University of Jyväskylä
Finland

ENRIC NART
Universitat Autònoma de Barcelona
Spain

MARCEL NICOLAU
Universitat Autònoma de Barcelona
Spain

MARC NOY
Universitat Politècnica de Catalunya
Spain

JOAQUÍN M. ORTEGA
Universitat Autònoma de Barcelona
Spain

AUGUST PALANQUES
Universitat de Barcelona
Spain

ATHENESE PAPADOPOULOS
Université de Strasbourg
France

CARLES PERELLÓ
Universitat Autònoma de Barcelona
Spain

DOLORS HERBERA
Universitat Autònoma de Barcelona
Spain

MASANORI ITAI
Saint Lawrence University, New York
USA

XAVIER JARQUE
Universitat Autònoma de Barcelona
Spain

VAUGHAN JONES
University of California, Berkeley
USA

PAULETTE LIBERMANN
Université de Paris
France

L. LÓPEZ BONILLA
Universitat de Barcelona
Spain

IRENE LLERENA
Universitat de Barcelona
Spain

ARMANDO MACHADO
Universidade de Lisboa
Portugal

FRANCISCO MARHUENDA
Universidad de Alicante
Spain

SEBASTIÀ MARTÍ
Universitat de Barcelona
Spain

MARTIN MATHIEU
Universität Tübingen
Germany

JAUME MONCASI
Universitat Autònoma de Barcelona
Spain

P. P. NARAYANASWAMI
Memorial University of Newfoundland
Canada

EMILIO NEGRÍN
Universidad de La Laguna
Spain

SERGEI NOVIKOV
Landau Institute, Moscow
Russia

MARIANO DEL OLMO
Universidad de Valladolid
Spain

MIGUEL A. PALACIOS
Universidad de Granada
Spain

VICENTE PALMER
Col·legi Universitari de Castelló
Spain

PERE PASCUAL
Universitat Politècnica de Catalunya
Spain

GEORGE PESCHKE
University of Alberta
Canada

MARKUS PFENNIGER
University of Wales
United Kingdom

JOAN PORTI
Universitat Autònoma de Barcelona
Spain

FERRAN PUERTA
Universitat Politècnica de Catalunya
Spain

RAFAEL ROBLES
Universidad de Sevilla
Spain

JORDI SALUDES
Universitat Politècnica de Catalunya
Spain

MARIANO SANTANDER
Universidad de Valladolid
Spain

JÜRGEN SCHWEIZER
Universität Tübingen
Germany

VLAD SERGIESCU
Institut Fourier, Grenoble
France

NIKOLAI SHCHERBINA
Institute of Mathematics, Kiev
Ukraine

STEPHEN SMALE
University of California, Berkeley
USA

IGNACIO SOLS
Universidad Complutense de Madrid
Spain

WIESLAW SZLENK
Uniwersytet Warszawski
Poland

RENÉ THOM
IHES
France

IZU VAISMAN
University of Haifa
Israel

ENRIC VENTURA
Universitat Autònoma de Barcelona
Spain

NÚRIA VILA
Universitat de Barcelona
Spain

JOSEP VIVES
Universitat Autònoma de Barcelona
Spain

JORGE WAGENSBERG
Universitat de Barcelona
Spain

KAI XU
University of Aberdeen
United Kingdom

MONTSERRAT PONS
Universitat Politècnica de Catalunya
Spain

JOSÉ PORTILLO
Universidad de Sevilla
Spain

AGUSTÍ REVENTÓS
Universitat Autònoma de Barcelona
Spain

CORA SADOSKY
Howard University
USA

MIGUEL SÁNCHEZ
Universidad de Granada
Spain

LAIA SAUMELL
Universitat Autònoma de Barcelona
Spain

SUDARSHAN K. SEHGAL
University of Alberta
Canada

FERRAN SERRANO
Universitat de Barcelona
Spain

ANTONI SINTES
Universitat Autònoma de Barcelona
Spain

LLUÍS SOLÉ
Universitat Politècnica de Catalunya
Spain

ARNE STRAY
University of Bergen
Norway

CRISTINA SZLENK
Uniwersytet Warszawski
Poland

VALERIO TOLEDANO
Ecole Polytechnique de Lausanne
Switzerland

STEPHANOS VENAKIDES
Duke University
USA

JOAN VERDERA
Universitat Autònoma de Barcelona
Spain

CORI VILELLA
Universitat Autònoma de Barcelona
Spain

JOSÉ LUIS VIVIENTE
Universidad de Zaragoza
Spain

GERALD E. WELTERS
Universitat de Barcelona
Spain

PAVOL ZLATOŠ
Univerzita Komenského, Bratislava
Czechoslovakia

Lecture Notes in Mathematics

For information about Vols. 1–1334
please contact your bookseller or Springer-Verlag

Vol. 1432: K. Ambos-Spies, G.H. Müller, G.E. Sacks (Eds.), Recursion Theory Week. Proceedings, 1989. VI, 393 pages. 1990.

Vol. 1433: S. Lang, W. Cherry, Topics in Nevanlinna Theory. II, 174 pages.1990.

Vol. 1434: K. Nagasaka, E. Fouvry (Eds.), Analytic Number Theory. Proceedings, 1988. VI, 218 pages. 1990.

Vol. 1435: St. Ruscheweyh, E.B. Saff, L.C. Salinas, R.S. Varga (Eds.), Computational Methods and Function Theory. Proceedings, 1989. VI, 211 pages. 1990.

Vol. 1436: S. Xambó-Descamps (Ed.), Enumerative Geometry. Proceedings, 1987. V, 303 pages. 1990.

Vol. 1437: H. Inassaridze (Ed.), K-theory and Homological Algebra. Seminar, 1987–88. V, 313 pages. 1990.

Vol. 1438: P.G. Lemarié (Ed.) Les Ondelettes en 1989. Seminar. IV, 212 pages. 1990.

Vol. 1439: E. Bujalance, J.J. Etayo, J.M. Gamboa, G. Gromadzki. Automorphism Groups of Compact Bordered Klein Surfaces: A Combinatorial Approach. XIII, 201 pages. 1990.

Vol. 1440: P. Latiolais (Ed.), Topology and Combinatorial Groups Theory. Seminar, 1985–1988. VI, 207 pages. 1990.

Vol. 1441: M. Coornaert, T. Delzant, A. Papadopoulos. Géométrie et théorie des groupes. X, 165 pages. 1990.

Vol. 1442: L. Accardi, M. von Waldenfels (Eds.), Quantum Probability and Applications V. Proceedings, 1988. VI, 413 pages. 1990.

Vol. 1443: K.H. Dovermann, R. Schultz, Equivariant Surgery Theories and Their Periodicity Properties. VI, 227 pages. 1990.

Vol. 1444: H. Korezlioglu, A.S. Ustunel (Eds.), Stochastic Analysis and Related Topics VI. Proceedings, 1988. V, 268 pages. 1990.

Vol. 1445: F. Schulz, Regularity Theory for Quasilinear Elliptic Systems and – Monge Ampère Equations in Two Dimensions. XV, 123 pages. 1990.

Vol. 1446: Methods of Nonconvex Analysis. Seminar, 1989. Editor: A. Cellina. V, 206 pages. 1990.

Vol. 1447: J.-G. Labesse, J. Schwermer (Eds), Cohomology of Arithmetic Groups and Automorphic Forms. Proceedings, 1989. V, 358 pages. 1990.

Vol. 1448: S.K. Jain, S.R. López-Permouth (Eds.), Non-Commutative Ring Theory. Proceedings, 1989. V, 166 pages. 1990.

Vol. 1449: W. Odyniec, G. Lewicki, Minimal Projections in Banach Spaces. VIII, 168 pages. 1990.

Vol. 1450: H. Fujita, T. Ikebe, S.T. Kuroda (Eds.), Functional-Analytic Methods for Partial Differential Equations. Proceedings, 1989. VII, 252 pages. 1990.

Vol. 1451: L. Alvarez-Gaumé, E. Arbarello, C. De Concini, N.J. Hitchin, Global Geometry and Mathematical Physics. Montecatini Terme 1988. Seminar. Editors: M. Francaviglia, F. Gherardelli. IX, 197 pages. 1990.

Vol. 1452: E. Hlawka, R.F. Tichy (Eds.), Number-Theoretic Analysis. Seminar, 1988–89. V, 220 pages. 1990.

Vol. 1453: Yu.G. Borisovich, Yu.E. Gliklikh (Eds.), Global Analysis – Studies and Applications IV. V, 320 pages. 1990.

Vol. 1454: F. Baldassari, S. Bosch, B. Dwork (Eds.), p-adic Analysis. Proceedings, 1989. V, 382 pages. 1990.

Vol. 1455: J.-P. Françoise, R. Roussarie (Eds.), Bifurcations of Planar Vector Fields. Proceedings, 1989. VI, 396 pages. 1990.

Vol. 1456: L.G. Kovács (Ed.), Groups – Canberra 1989. Proceedings. XII, 198 pages. 1990.

Vol. 1457: O. Axelsson, L.Yu. Kolotilina (Eds.), Preconditioned Conjugate Gradient Methods. Proceedings, 1989. V, 196 pages. 1990.

Vol. 1458: R. Schaaf, Global Solution Branches of Two Point Boundary Value Problems. XIX, 141 pages. 1990.

Vol. 1459: D. Tiba, Optimal Control of Nonsmooth Distributed Parameter Systems. VII, 159 pages. 1990.

Vol. 1460: G. Toscani, V. Boffi, S. Rionero (Eds.), Mathematical Aspects of Fluid Plasma Dynamics. Proceedings, 1988. V, 221 pages. 1991.

Vol. 1461: R. Gorenflo, S. Vessella, Abel Integral Equations. VII, 215 pages. 1991.

Vol. 1462: D. Mond, J. Montaldi (Eds.), Singularity Theory and its Applications. Warwick 1989, Part I. VIII, 405 pages. 1991.

Vol. 1463: R. Roberts, I. Stewart (Eds.), Singularity Theory and its Applications. Warwick 1989, Part II. VIII, 322 pages. 1991.

Vol. 1464: D. L. Burkholder, E. Pardoux, A. Sznitman, Ecole d'Eté de Probabilités de Saint- Flour XIX-1989. Editor: P. L. Hennequin. VI, 256 pages. 1991.

Vol. 1465: G. David, Wavelets and Singular Integrals on Curves and Surfaces. X, 107 pages. 1991.

Vol. 1466: W. Banaszczyk, Additive Subgroups of Topological Vector Spaces. VII, 178 pages. 1991.

Vol. 1467: W. M. Schmidt, Diophantine Approximations and Diophantine Equations. VIII, 217 pages. 1991.

Vol. 1468: J. Noguchi, T. Ohsawa (Eds.), Prospects in Complex Geometry. Proceedings, 1989. VII, 421 pages. 1991.

Vol. 1469: J. Lindenstrauss, V. D. Milman (Eds.), Geometric Aspects of Functional Analysis. Seminar 1989-90. XI, 191 pages. 1991.

Vol. 1470: E. Odell, H. Rosenthal (Eds.), Functional Analysis. Proceedings, 1987-89. VII, 199 pages. 1991.

Vol. 1471: A. A. Panchishkin, Non-Archimedean L-Functions of Siegel and Hilbert Modular Forms. VII, 157 pages. 1991.

Vol. 1472: T. T. Nielsen, Bose Algebras: The Complex and Real Wave Representations. V, 132 pages. 1991.

Vol. 1473: Y. Hino, S. Murakami, T. Naito, Functional Differential Equations with Infinite Delay. X, 317 pages. 1991.

Vol. 1474: S. Jackowski, B. Oliver, K. Pawałowski (Eds.), Algebraic Topology, Poznań 1989. Proceedings. VIII, 397 pages. 1991.

Vol. 1475: S. Busenberg, M. Martelli (Eds.), Delay Differential Equations and Dynamical Systems. Proceedings, 1990. VIII, 249 pages. 1991.

Vol. 1476: M. Bekkali, Topics in Set Theory. VII, 120 pages. 1991.

Vol. 1477: R. Jajte, Strong Limit Theorems in Noncommutative L_2-Spaces. X, 113 pages. 1991.

Vol. 1478: M.-P. Malliavin (Ed.), Topics in Invariant Theory. Seminar 1989-1990. VI, 272 pages. 1991.

Vol. 1479: S. Bloch, I. Dolgachev, W. Fulton (Eds.), Algebraic Geometry. Proceedings, 1989. VII, 300 pages. 1991.

Vol. 1480: F. Dumortier, R. Roussarie, J. Sotomayor, H. Żołądek, Bifurcations of Planar Vector Fields: Nilpotent Singularities and Abelian Integrals. VIII, 226 pages. 1991.

Vol. 1481: D. Ferus, U. Pinkall, U. Simon, B. Wegner (Eds.), Global Differential Geometry and Global Analysis. Proceedings, 1991. VIII, 283 pages. 1991.

Vol. 1482: J. Chabrowski, The Dirichlet Problem with L^2-Boundary Data for Elliptic Linear Equations. VI, 173 pages. 1991.

Vol. 1483: E. Reithmeier, Periodic Solutions of Nonlinear Dynamical Systems. VI, 171 pages. 1991.

Vol. 1484: H. Delfs, Homology of Locally Semialgebraic Spaces. IX, 136 pages. 1991.

Vol. 1485: J. Azéma, P. A. Meyer, M. Yor (Eds.), Séminaire de Probabilités XXV. VIII, 440 pages. 1991.

Vol. 1486: L. Arnold, H. Crauel, J.-P. Eckmann (Eds.), Lyapunov Exponents. Proceedings, 1990. VIII, 365 pages. 1991.

Vol. 1487: E. Freitag, Singular Modular Forms and Theta Relations. VI, 172 pages. 1991.

Vol. 1488: A. Carboni, M. C. Pedicchio, G. Rosolini (Eds.), Category Theory. Proceedings, 1990. VII, 494 pages. 1991.

Vol. 1489: A. Mielke, Hamiltonian and Lagrangian Flows on Center Manifolds. X, 140 pages. 1991.

Vol. 1490: K. Metsch, Linear Spaces with Few Lines. XIII, 196 pages. 1991.

Vol. 1491: E. Lluis-Puebla, J.-L. Loday, H. Gillet, C. Soulé, V. Snaith, Higher Algebraic K-Theory: an overview. IX, 164 pages. 1992.

Vol. 1492: K. R. Wicks, Fractals and Hyperspaces. VIII, 168 pages. 1991.

Vol. 1493: E. Benoît (Ed.), Dynamic Bifurcations. Proceedings, Luminy 1990. VII, 219 pages. 1991.

Vol. 1494: M.-T. Cheng, X.-W. Zhou, D.-G. Deng (Eds.), Harmonic Analysis. Proceedings, 1988. IX, 226 pages. 1991.

Vol. 1495: J. M. Bony, G. Grubb, L. Hörmander, H. Komatsu, J. Sjöstrand, Microlocal Analysis and Applications. Montecatini Terme, 1989. Editors: L. Cattabriga, L. Rodino. VII, 349 pages. 1991.

Vol. 1496: C. Foias, B. Francis, J. W. Helton, H. Kwakernaak, J. B. Pearson, H_∞-Control Theory. Como, 1990. Editors: E. Mosca, L. Pandolfi. VII, 336 pages. 1991.

Vol. 1497: G. T. Herman, A. K. Louis, F. Natterer (Eds.), Mathematical Methods in Tomography. Proceedings 1990. X, 268 pages. 1991.

Vol. 1498: R. Lang, Spectral Theory of Random Schrödinger Operators. X, 125 pages. 1991.

Vol. 1499: K. Taira, Boundary Value Problems and Markov Processes. IX, 132 pages. 1991.

Vol. 1500: J.-P. Serre, Lie Algebras and Lie Groups. VII, 168 pages. 1992.

Vol. 1501: A. De Masi, E. Presutti, Mathematical Methods for Hydrodynamic Limits. IX, 196 pages. 1991.

Vol. 1502: C. Simpson, Asymptotic Behavior of Monodromy. V, 139 pages. 1991.

Vol. 1503: S. Shokranian, The Selberg-Arthur Trace Formula (Lectures by J. Arthur). VII, 97 pages. 1991.

Vol. 1504: J. Cheeger, M. Gromov, C. Okonek, P. Pansu, Geometric Topology: Recent Developments. Editors: P. de Bartolomeis, F. Tricerri. VII, 197 pages. 1991.

Vol. 1505: K. Kajitani, T. Nishitani, The Hyperbolic Cauchy Problem. VII, 168 pages. 1991.

Vol. 1506: A. Buium, Differential Algebraic Groups of Finite Dimension. XV, 145 pages. 1992.

Vol. 1507: K. Hulek, T. Peternell, M. Schneider, F.-O. Schreyer (Eds.), Complex Algebraic Varieties. Proceedings, 1990. VII, 179 pages. 1992.

Vol. 1508: M. Vuorinen (Ed.), Quasiconformal Space Mappings. A Collection of Surveys 1960-1990. IX, 148 pages. 1992.

Vol. 1509: J. Aguadé, M. Castellet, F. R. Cohen (Eds.), Algebraic Topology - Homotopy and Group Cohomology. Proceedings, 1990. X, 330 pages. 1992.

Vol. 1510: P. P. Kulish (Ed.), Quantum Groups. Proceedings, 1990. XII, 398 pages. 1992.

Vol. 1511: B. S. Yadav, D. Singh (Eds.), Functional Analysis and Operator Theory. Proceedings, 1990. VIII, 223 pages. 1992.

Vol. 1512: L. M. Adleman, M.-D. A. Huang, Primality Testing and Abelian Varieties Over Finite Fields. VII, 142 pages. 1992.

Vol. 1513: L. S. Block, W. A. Coppel, Dynamics in One Dimension. VIII, 249 pages. 1992.

Vol. 1514: U. Krengel, K. Richter, V. Warstat (Eds.), Ergodic Theory and Related Topics III, Proceedings, 1990. VIII, 236 pages. 1992.

Vol. 1515: E. Ballico, F. Catanese, C. Ciliberto (Eds.), Classification of Irregular Varieties. Proceedings, 1990. VII, 149 pages. 1992.

Vol. 1516: R. A. Lorentz, Multivariate Birkhoff Interpolation. IX, 192 pages. 1992.

Vol. 1517: K. Keimel, W. Roth, Ordered Cones and Approximation. VI, 134 pages. 1992.

Vol. 1518: H. Stichtenoth, M. A. Tsfasman (Eds.), Coding Theory and Algebraic Geometry. Proceedings, 1991. VIII, 223 pages. 1992.

Vol. 1519: M. W. Short, The Primitive Soluble Permutation Groups of Degree less than 256. IX, 145 pages. 1992.

Vol. 1520: Yu. G. Borisovich, Yu. E. Gliklikh (Eds.), Global Analysis - Studies and Applications V. VII, 284 pages. 1992.

Vol. 1521: S. Busenberg, B. Forte, H. K. Kuiken, Mathematical Modelling of Industrial Process. Bari, 1990. Editors: V. Capasso, A. Fasano. VII, 162 pages. 1992.

Vol. 1522: J.-M. Delort, F. B. I. Transformation. VII, 101 pages. 1992.

Vol. 1523: W. Xue, Rings with Morita Duality. X, 168 pages. 1992.

Vol. 1524: M. Coste, L. Mahé, M.-F. Roy (Eds.), Real Algebraic Geometry. Proceedings, 1991. VIII, 418 pages. 1992.

Vol. 1525: C. Casacuberta, M. Castellet (Eds.), Mathematical Research Today and Tomorrow. VII, 112 pages. 1992.

4. Lecture Notes are printed by photo-offset from the master-copy delivered in camera-ready form by the authors. Springer-Verlag provides technical instructions for the preparation of manuscripts. Macro packages in T_EX, L^AT_EX2e, $L^AT_EX2.09$ are available from Springer's web-pages at

http://www.springer.de/math/authors/b-tex.html.

Careful preparation of the manuscripts will help keep production time short and ensure satisfactory appearance of the finished book.

The actual production of a Lecture Notes volume takes approximately 12 weeks.

5. Authors receive a total of 50 free copies of their volume, but no royalties. They are entitled to a discount of 33.3 % on the price of Springer books purchase for their personal use, if ordering directly from Springer-Verlag.

Commitment to publish is made by letter of intent rather than by signing a formal contract. Springer-Verlag secures the copyright for each volume. Authors are free to reuse material contained in their LNM volumes in later publications: A brief written (or e-mail) request for formal permission is sufficient.

Addresses:

Professor F. Takens, Mathematisch Instituut,
Rijksuniversiteit Groningen, Postbus 800,
9700 AV Groningen, The Netherlands
E-mail: F.Takens@math.rug.nl

Professor B. Teissier
Université Paris 7
UFR de Mathématiques
Equipe Géométrie et Dynamique
Case 7012
2 place Jussieu
75251 Paris Cedex 05
E-mail: Teissier@math.jussieu.fr

Springer-Verlag, Mathematics Editorial, Tiergartenstr. 17,
D-69121 Heidelberg, Germany,
Tel.: *49 (6221) 487-701
Fax: *49 (6221) 487-355
E-mail: lnm@Springer.de